ANNE HESSEL, JEAN JOUZEL AND
PIERRE LARROUTUROU

A CLIMATE PACT FOR EUROPE

How to Finance the Green Deal

First published in Great Britain in 2021 by

Bristol University Press
University of Bristol
1–9 Old Park Hill
Bristol
BS2 8BB
UK
t: +44 (0)117 954 5940
e: bup-info@bristol.ac.uk

Details of international sales and distribution partners are available at
bristoluniversitypress.co.uk

British Library Cataloguing in Publication Data
A catalogue record for this book is available from the British Library

ISBN 978-1-5292-1913-5 paperback
ISBN 978-1-5292-1914-2 ePub
ISBN 978-1-5292-1915-9 ePdf

Cover design: Liam Roberts
Front cover image: LightFieldStudios
Bristol University Press uses environmentally responsible print partners.
Printed and bound in Great Britain by CMP, Poole

Contents

List of Tables, Figures and Boxes

Tables

Figures

Boxes

Acknowledgements: The English Translation

It has been an education for me to translate the inspiring French version written in 2018. I share the commitment and energy of the authors, if not all the content, and trust this book will help us to win the battle against deregulation of our climate – as the IPCC says in 2021, we have only one decade to do this. Being neither very literate, nor having translated any other important work, I am hugely grateful to Martin Phillips, who has improved all the chapters, and to Deborah Honoré, who made my translation of Chapter Three much clearer and succeeded in making parts of it beautiful to read. Warm thanks to Alice Pradel, who translated Chapter Five, and Blandine Cheval, who read chapters in French for correction. As usual, the greatest help came from Alison Duncan, who launched this whole adventure, provided ideas on strategic partnerships and read most of the French book, while I checked for errors of omission in the English, and then the translation, often in dark hours of the night.

Patrick Duncan, August 2019

How We Can Win the Battle

Nicolas Hulot[1]

In an emergency, when it comes to funding, if you do not look outside the box, you won't make it. What if Europe decided (finally!) to put finance at the service of the climate? We need a new economic model that is both sustainable and inclusive. We must convince the people of Europe that this new economic model is desirable and beneficial for everyone, and that it brings new protections and new economic opportunities.

To win this battle, you have to win the battle of ideas. It is the drama of this time that short-term profits are incompatible with the protection of the socio-ecosystems we live in, and incompatible with freeing people. It's the 'tragedy of the horizon', as the Governor of the Bank of England, Mark Carney, puts it.

Europe must clearly lay the foundations for a new economic and financial model – this is vital for humanity. And for Europe, this is a wonderful opportunity for her to reconnect with her citizens.

15 March 2018 at the United Nations Educational, Scientific and Cultural Organization (UNESCO), Paris

2020: A Warning Shot

Anne Hessel

The year 2020 will have been one of all-out experiments to try to save our skins from a deadly risk that spread in a few weeks – COVID-19! This crisis has revealed the fragility of our economies, which have not been able to maintain the production of life's essentials locally – food, medicines and so on.

This crisis has also obliged us to experience lives of low consumption and no travel, which many have found rather pleasant! This crisis makes us fear the next: indeed, epidemics can arise from thawing permafrost, which is happening too quickly and reactivating antediluvian viruses. Finally, this kind of crisis will be totally destructive in the event of a similar epidemic under a heatwave!

This warning shot convinced many of our fellow people to act against the deregulation of our climate, but there are still those who doubt the reality of global warming: some realize that warming is happening but are not aware of the gravity of what is on the way; others know that it is serious but do not wish to address the problem. Although written before, the COVID-19 pandemic makes this book highly topical since the book highlights the consequences of the economic globalization developed at a brisk pace since the early 1980s, as well as its fragility, with near-monopolies of production concentrated in certain countries. Being on the other side of the world, these monopolies deprive many people of equipment to react to the pandemic – objects as simple as surgical masks or gloves.

Faced with the urgent need to restart the economy, some dream of returning to the 'world before', making the pandemic a mere parenthesis. Others want to take the opposite view and preach the relocation of production, even the return to economic sovereignty. The closure of national borders during the health crisis has shown, they say, that the only community that can stand up in a crisis is the good old *national* community inherited from the past. These two apparently opposite attitudes share the confusion between an eminently reversible economic globalization and the globalization of interdependencies between humanity (and its societies) and the biosphere, which is irreversible.

The speeches on relocation and sovereignty reinforce the nationalist counter-reform, embodied by Donald Trump, ruining the efforts of multilateralism and reinventing 'the enemy from abroad' at the very moment that the pandemic is embodying the breadth of global interdependencies and the inability of a nation to cope with the disease through national decisions. The pandemic poses, on a planetary scale, the major question of responsibility, from the responsibilities of each of us – wearing a mask to protect others – to the responsibility of the various private and state actors in the emergence and spread of the virus. The pandemic reminds us that good health for all should be a global, common objective.

The parallel with the integrity of the climate is clear. The integrity of the climate requires correcting decades of energy-consuming buildings, energy-consuming vehicles, energy-consuming businesses, agricultural systems that produce too much of the greenhouse gases and our individual habits as regards eating, warming and air conditioning.

This book aims to restore the integrity of the climate through a major relaunch of the real economy. By doing so, it will lead the way to a more serene climate, both in politics and in the weather, that better protects us from subsequent pandemics. Doctors understand that Europe was on the brink of a catastrophe from which our societies would not have survived: the COVID-19 pandemic combined with a heatwave like that of 2003 or 2019. In such a case, the confinements would have been made impossible, and the pandemic could have killed perhaps a quarter of the European population by quickly saturating the hospitals.

The health crisis we have just experienced brings an inspiring idea: the measures taken to stop the COVID-19 pandemic demonstrate that it is realistic to rapidly make political decisions to allocate huge loans and a substantial budget to defeat a scourge in Europe. It is possible; we are doing it! In a few weeks ... This proves that the solutions proposed in this book (an annual loan of €1,000 billion and a European budget of €100 billion per year) are realistic and acceptable political solutions should the urgency of climate change be perceived as that of the health crisis. This aspect – 'Just do it. It is easy. It is a political decision' – is the overwhelming lesson of the COVID-19 crisis – a lesson that enchants us, and one that would have enchanted the incurable optimist Stéphane Hessel.

Foreword

Mary Robinson

This is a well-researched, readable and very timely book. It was written in a different era, before the COVID-19 pandemic. Now, we have the benefit of the lessons of COVID-19 and how important they are to the arguments for the Climate Pact:

- We have learned that human behaviour matters – it is our collective compliance with physical distancing that is keeping the virus in check.
- We have learned that science matters, as politicians are guided in policymaking by health experts – as they should be by climate scientists.
- We have learned that when economies are forced to freeze, enormous sums of money become available, not only to seek medical solutions such as protective equipment and vaccines, but also to furlough and subsidize workers so that this will make an economic recovery more rapid.
- We have learned that government and leadership matters, as failures and delays in dealing with COVID-19 are cruelly exposed in rising death levels.

These lessons, I believe, equip us now about how to finance the Green Deal as we start to emerge from the COVID-19 crisis. We need to do this in a way that is compatible with the other looming crisis – the climate crisis. I was struck by a sentence at the beginning of Chapter Seven, which considers the possibility of €1,000 billion for the climate and asks whether we are ready: 'For the moment, in all our countries, we meet a "financial cliff" that prevents the best wills in the world from deploying actions at the level required.' That cliff is now gone! Countries and international institutions are breaking all the normal rules to cope with the economic crisis caused by COVID-19, and nothing will ever be the same again.

So, the question is: how do we 'build back better'? It is not automatic that we will do so in a way that addresses the climate crisis. China, the first country to be affected by COVID-19 and the first to begin the recovery, is planning more new coal plants than ever in order to fuel its recovery.

This is the challenge for Europe: to build on the €1 trillion European Union Recovery Fund that has been announced and make sure that it becomes a true Climate Pact, as advocated in this book. I strongly support putting finance back to the service of the common good with ideas like the European Climate Finance Pact and a European Climate Bank. It seems reasonable to stop European fiscal dumping by looking for a contribution of 5 per cent of corporate profits that are not to be reinvested in Europe. It would also be necessary to adopt the other policies advocated, such as directing private financing to green investment, stopping all subsidies of fossil fuel and increasing the price of carbon.

I was glad to see Stéphane Hessel, a man I loved and admired hugely, quoted in the Conclusion as follows: '"We experienced apartheid and the end of apartheid", he said: "We had the Berlin wall and saw the end of the wall.... It's up to us, the citizens, to decide on our future."' So, let me add my own words to Stéphane's: we are experiencing COVID-19, and emerging from it to a just and inclusive economy, sustainable with nature, is our only way to ensure a future for our children and grandchildren.

ONE

'Our Home Is Burning and We Are Looking Elsewhere'

> Our home is burning and we are looking elsewhere. Nature, mutilated, overexploited, no longer manages to restore herself and we refuse to admit it. Humanity suffers. It suffers from poor development, both in the North and the South, and we are indifferent.
>
> Earth and humanity are in danger and we are all responsible for it.
>
> It's time to open our eyes. On all continents, the warning signals are clear.... We cannot say we did not know! We need to take care that the twenty-first century does not become, for future generations, that of a crime of humanity against life.
>
> Jacques Chirac, Johannesburg, 2002

When Jacques Chirac pronounced these words at the World Sustainable Development Summit in Johannesburg on 2 September 2002, some still had doubts about the reality of global warming. The image of the house that is burning and 'a crime of humanity against life' may have seemed excessive to them.

Today, no one can seriously doubt the reality of our climatic disruption. The last six years (2015 to 2020) were the hottest years since scientists have measured global temperatures, that is, since the mid-19th century, and each year, several tens of thousands of homes are destroyed by extreme weather events (see Figure 1.1).

Figure 1.1: Variations in global mean temperature

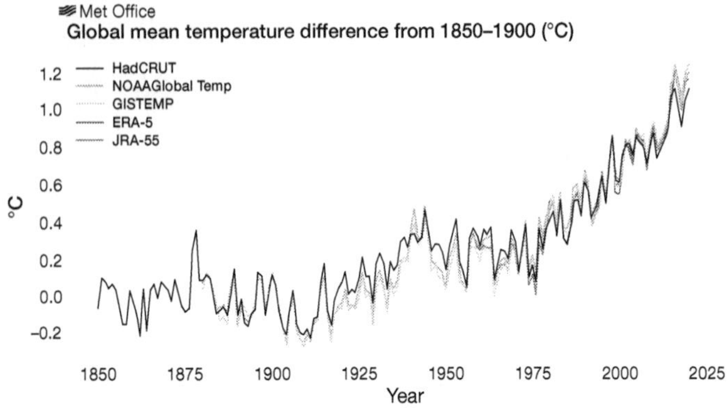

Source: © Crown Copyright 2021. Information provided by the National Meteorological Library and Archive – Met Office, UK.

Not a month goes by without hundreds of thousands of people across the planet seeing their lives turned upside down by droughts, heatwaves, forest fires or, at the other extreme, torrential rains and floods. In just a few decades, global warming has been accompanied by disruptions of water cycles.

In Japan, in July 2018, rain and floods caused more than 200 deaths, and tens of thousands of homes were evacuated. At the same time, in Canada, the heatwave caused 50 deaths. Two weeks later, Japan was again hit by a deadly heatwave, then it was California that had to fight against the worst forest fires in its history.

Meanwhile, Europe was undergoing an extraordinary heatwave. Forest fires were affecting Sweden and Greece at the same time, causing dozens of deaths near Athens. In Portugal, memories of the fires of the summer of 2017, which killed more than 110 people, were revived. In the Netherlands, some sections of motorways had to be closed because the asphalt was melting. Many parts of Germany were suffering from drought, and in a few weeks, the flow of the rivers had fallen sharply: 'The Rhine is difficult to navigate and in some parts of the Elbe there are less than 30 centimetres of water.' In 2017, it had been the Czech Republic and Northern Hungary that suffered a similar drought.

In France, fortunately, the heatwave of 2018 was much less serious than that of 2003, but it will remain a very painful moment for millions of us (How does one live normally when it is too hot, even in the middle of the night? How does one get to sleep?). In Eastern France, the drought was so bad that at the beginning of August, the pastures were dried out and the farmers had to feed cattle and sheep the fodder reserves saved for the winter.

The heatwave forced Electricité de France to shut down four nuclear power stations: "Last year the average flow of the Rhone was 30 per cent below that of a normal year", explained a senior executive of the Compagnie Nationale du Rhône in 2018:

> 'We had to discreetly ask our Swiss friends to release water to cool a nuclear power station. This year, we had to shut down four power stations because the water in the rivers, which should cool them, was too warm.... What will happen in 20 years if the warming continues? Our system is without doubt more fragile than we think.'

Fragile, our system? Probably, yes, much more so than we think! Due to the heatwave that hit much of the planet, crop yields in 2018 were lower than expected, and the price of cereals began to increase sharply in the middle of the summer of 2018. This reminded specialists of how major floods hit the Paris Basin in June 2016 and caused serious crop damage: with their roots in water, the ears of wheat grains did not fill properly; some even began to rot before harvest. In France and Belgium, the wheat harvest was 31 per cent below the average of the previous decade. A loss of 31 per cent is considerable!

Fortunately, the other wheat granary in Europe had a very good harvest that year. However, if the Ukraine had also been affected by floods (or by a heatwave leading to a serious drop in yields), we might have received ration tickets to manage the shortage that these events would have triggered.

With the internet, it is often enough to simply click to get goods or services while sitting comfortably in your living room, but we must not forget that we are all mammals, accustomed to two or three meals per day. Moreover, the grain stores represent less than a third of our annual consumption. What would happen if for two years running, the harvests in the two European wheat granaries were down by 30 or 40 per cent? We know that food riots have occurred in Africa and South America when the price of cereals became too high for some people to afford; it is a mistake to think that Europe and the other rich countries will be safe from this kind of trouble for ever.

'Our house is burning …' Many of us thought about Jacques Chirac's speech in December 2017 when giant forest fires devastated California, mobilizing more than 5,000 firefighters and leading to the evacuation of 300,000 residents. Donald Trump was looking elsewhere, but more than a dozen people died and many thousands of homes were destroyed by this new type of fire – of course, there have always been forest fires in California, as in most of the forested areas of the world. These are usually triggered by heat in the summer. If fires break out in December, if they are so violent and if it is necessary to mobilize several thousands of firefighters to fight them, then it means that the forests have not managed to recover from California's five years of drought.

Heatwaves, droughts, fires, torrential rains, floods and hurricanes – according to insurance companies, the number of extreme weather events has more than doubled in less than 30 years (see Figure 1.2). Although people and goods are more exposed to risk, and this is one important cause[1] of the increases in costs, the costs are more than impressive, especially since Figure 1.2 is likely to underestimate the costs of the extraordinary climatic events that have affected the countries of the southern hemisphere – where many citizens are

Figure 1.2: The costs of catastrophes linked to climate change

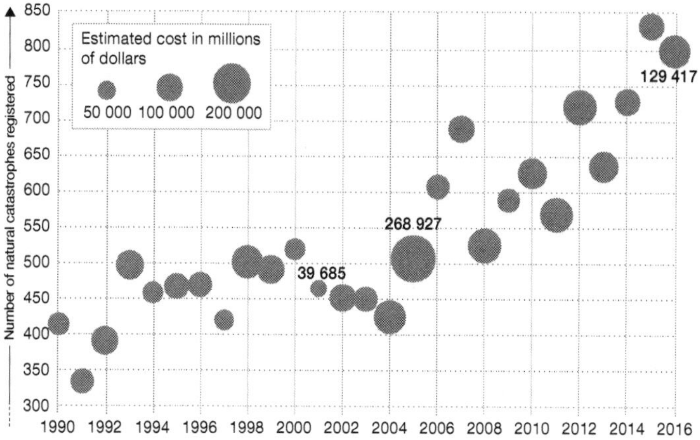

Source: *The Lancet Countdown* from AFP.

poorly insured or not insured at all. What is more, these data stop in 2016. We know now that in 2017 in the US alone, the damage caused by climatic events such as hurricanes and forest fires cost more than US$307 billion[2] and that 2018 was also marked by many disasters!

Some events have remained etched in our memories, like Hurricane Katrina, which devastated New Orleans on 29 August 2005, causing the deaths of more than 1,800 people, both young and old, trapped in their homes or washed away as they tried to flee the waves, with pieces of sheet metal carried away by the wind and trees pulled over. This hurricane forced over a million people to evacuate the area.

The number of hurricanes has not increased, but in the North Atlantic, the proportion of very violent hurricanes (with winds exceeding 200 km/h) has doubled in 40 years. There has also been a large number of less publicized events that also precipitated terrible dramas; for example, the torrential rains that caused 75 deaths and injured thousands of people in Lima, Peru, in March 2017 or the floods in India, Nepal and

Bangladesh, which left more than 200 dead and caused 200,000 people to be evacuated in August 2017. There are the same images of desolation each time: rains of incredible violence; houses carried away by avalanches of mud; cars washed away by the waters; and the strongest people trying to protect the young and old, though no one is safe.

Furthermore, Peru, India, Nepal and Bangladesh are obviously not the only places affected. In Kenya, in April 2018, people had been waiting for the rains for weeks, and when the rain arrived, it was so violent that about 100 people were killed and 200,000 were forced to leave their homes: 'the evacuation was made very difficult because the main roads and bridges were destroyed by the wild waters'.[3] Closer to home, in Nice on the Côte d'Azur, in October 2015, it took just a few hours for torrential rains to kill 20 people and cause chaos.

On all continents, water causes despair when it arrives too brutally and too violently, and this happens more and more frequently. However, there is also despair when the water runs out. These repeated and longer droughts obviously have very serious consequences in rural areas, but the lack of water now affects large cities, where we thought we were safe – we have since learnt that this is no longer the case.

In Bolivia, La Paz ran out of water in 2015. In Italy, in August 2017, it was the inhabitants of Rome and surrounding municipalities who had to limit their water consumption, and at the beginning of 2018, the people of the Cape in South Africa were severely affected:

> South Africa's second largest city, Cape Town has been hit by its worst drought in a century. The water reserves are at their lowest, to a point where the taps could soon run dry. To delay the arrival of this day-without-water, the 4 million inhabitants of the city were encouraged to use only 50 litres of water per person per day, the equivalent of a three minute shower. Local authorities have threatened to impose fines on those who exceed

> this threshold. If 'zero day' occurs, Capetonians will have to get their water from 200 water points, where they will receive a maximum of 25 litres per day per person.[4]

What will happen in 20 or 30 years if the warming continues? So far, Europe has been relatively unaffected by climate change, but researchers from Météo-France and the Pierre Simon Laplace Institute have announced that heatwaves are becoming more frequent. The summer of 2003 will become the norm after 2050, and every two or three years, we could experience heatwaves up to twice as hot[5] and over longer and longer periods: the season

---

Box 1.1: Ethiopia: the worst drought in 30 years

Ethiopia is in dire need. The drought that hit the country for several months destroyed crops. The government estimates that 10 million Ethiopians risk starvation.

Whole fields of crops have burnt off. The tomatoes suffered in the heat and, most important, did not get enough water. There was almost no rain from June to September and the crops were lost. Abu Dogeda, farmer: 'There is nothing at all. All this year's maize is like that, it's not edible. Even for cattle, if they eat that, it will kill them.'

Livestock, too, has suffered from drought, the worst in thirty years according to the United Nations. Demedawe Lolu, farmer: 'The goats have no water to drink. It may be one to three months that the livestock has not drunk water.'

To help its population, mostly farmers, the Ethiopian government has imported one million tonnes of wheat and spent 175 million euros on aid, but for the United Nations, twice as much is needed. This farmer agrees: 'The help we received is not enough to feed the eight members of the family for a fortnight. We are already exposed to famine. Unless the government takes further action and distributes aid, we will end up starving and die of hunger.'

TV5 Monde, 29 May 2017.

---

of heatwaves could begin in June and end in late September or mid-October. This obviously changes their potential impact: if the heatwaves occur in August, their consequences for crops are much less severe (since harvesting is completed) than if they take place in June or early July, when the grains are filling. Similarly, if heatwaves occur in August, they are less tiring for a large part of the population, that is, those on holiday. In June, millions of workers have to go to work even if they have slept badly, and they are in traffic jams or on public transport when the heat rises. To suffer a heatwave in a crowded metro or in stationary traffic is particularly difficult – exhausting.

According to simulations carried out with a weather model of Météo-France, if nothing radical is done to fight against the current warming, at the end of the century, we could suffer record temperatures of the order of 50 °C from time to time, even approaching 55 °C in some regions (see Figure 1.3).[6] When we see the difficulties encountered by many of us in dealing with heatwaves at 39 °C or 40 °C today (only two or three degrees above the 37 °C of our normal body temperature), it is better not to imagine the effects of living at 50 °C or 55 °C. "It's the boxer syndrome", says a Météo-France researcher:

> 'The boxer can cope with a huge blow on the head: he falls then gets up. He takes a second blow, falls and then he gets up. But the third blow can be fatal.... We cannot yet explain why, but living bodies – be they human, animal or forests – seem to react similarly. They cope with one heatwave, then a second, but the third risks being fatal.'

In this scenario – which is the 'central scenario' only if we continue to do nothing serious about global warming – mortality is likely to increase sharply: in the second half of the century, heatwaves could cause 50 times more deaths than currently. Today, heatwaves cause some 3,000 deaths every

Figure 1.3: Heatwaves could become more frequent in France, reaching 50°C

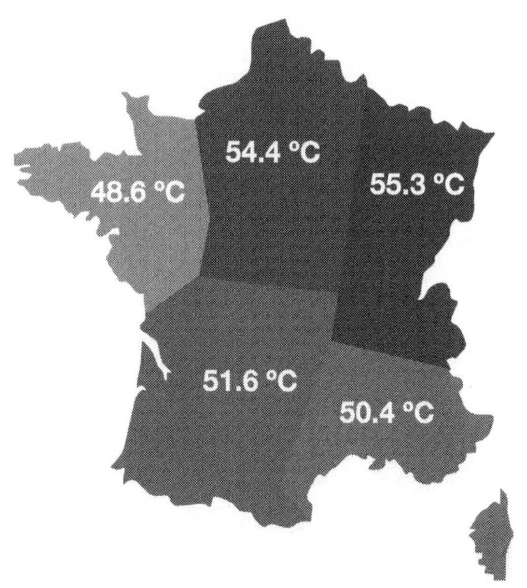

year in Europe. At the end of the century, they could cause 150,000 deaths each year.

We have already mentioned the impact of climate change on crop harvests. This question will become more and more crucial on all continents if nothing is done to combat the ongoing global warming.

In Africa, harvests may be halved as the population doubles

In Africa, the Food and Agriculture Organization's (FAO's) central scenario predicts crop losses of 35–60 per cent, depending on the region (and the level of global warming). In many regions of Africa, water will become scarcer but will fall more violently[7], as we saw earlier, while evaporation will increase. If nothing decisive is done in the coming years, the climate models

Table 1.1: Harvests halved?

Global warming	1.5 °C WORLD	2 °C WORLD
Expected wheat crop losses		
Western Africa	45% loss	60% loss
Eastern Africa	25% loss	35% loss
Central America	25% loss	40% loss

Source: Climate Analytica.

suggest that we will not avoid global warming above 2–3 °C. This would lead to harvests halved or more, while by 2050, the population of Africa is expected to double (see Table 1.1).

One has to be blind not to understand the dramatic effects on people who live in Africa if the climate chaos continues to get worse. Furthermore, Africa is not the only part of the world for which the scenarios appear more and more catastrophic if nothing is done to stop global warming. Some countries could become completely uninhabitable by 2070.

To keep our brain, our liver and all our organs working, they need to stay at about 37 °C. As outside temperatures rise, we evaporate water by sweating. This system of cooling functions well, but if the air outside is too hot and too humid, it does not. Our cooling system then stops working, and the human body has the greatest difficulty in surviving.

We have already seen waves of 'suffocating humid heat' in some parts of the world and for short periods; when it is too hot and humid, day and night, there is nothing else to do but to seek coolness! During the summer of 2017, one saw on social networks disturbing videos showing Chinese people leaving their homes at night to throw themselves on the grass under trees, begging for a bit of coolness.

Fortunately, in 2017, this humid heatwave lasted only two or three days. However, forecasts by researchers from the Massachusetts Institute of Technology (MIT)[8] show that if nothing effective is done against climate change, long and recurrent periods of stifling humid heat could make it very difficult to inhabit some of Asia's most populated areas in only 50 years.

In Africa, harvests halved while the population doubles by 2050? In Asia, a region inhabited by 400 million people in

Box 1.2: Part of China uninhabitable by 2070

According to the Massachusetts Institute of Technology, the northern plains of China – which have the greatest concentration of humans on the planet – could become the place the most prone to heatwaves on earth, by 2070. So much so that the 400 million residents of the province could no longer live there.

'This region is going to be the most affected by deadly humid heatwaves in the future,' says MIT Professor Elfatig Eltahir, who led the study. Humid heatwaves, which could occur repeatedly by the end of the century, would be so violent that they would kill people in good health in just six hours.

Already hard hit by climate change, Bangladesh risks being literally overwhelmed by the rising waters: according to an OECD [Organization for Economic Co-operation and Development] study, in Dacca, the capital, more than 11 million people could be exposed to dramatic floods by 2070, due to rising sea levels. Stifling heatwaves also threaten Bangladesh, Pakistan and parts of India: 'Increasing temperatures and humidity in the summer could reach levels beyond the human body's ability to survive unprotected', MIT researchers said in 2017.

In 2015, the same MIT team drew a similar conclusion for the Gulf countries. Based on regional climate simulations, the researchers estimated that 'wet weather peaks in the Gulf region are likely to approach and exceed this critical threshold' if greenhouse gas emissions remain on their current trajectory.

Sipos, A. (2018) *Le Parisien*, 2 August.

which survival could become very difficult by the end of the century? Cities of 10 million inhabitants that could be partially submerged by the waves?

If one takes the time to read carefully the more and more detailed studies published in recent years, there is no longer any doubt about the seriousness of what awaits us if we do not take action now. *So, how can one explain the unbelievable weakness of our reactions?*

Certainly, Europe has been relatively protected so far, and once the memories of the summer heatwaves pass, it is still possible for some leaders to 'look elsewhere', deny the problem and return to *business as usual.* However, by multiplying the extraordinary climatic events, Mother Nature is sending us very clear signals to help us imagine what awaits us in 20 or 30 years, more and more strongly, and more and more unpredictably, if we do not act very quickly.

It is no coincidence that the Intergovernmental Panel on Climate Change (IPCC) received the Nobel Peace Prize (and not a Nobel Prize in Chemistry or Physics): it is world peace that is at stake. It is sometimes said that we must 'save the planet', but the truth is that the planet will recover very well from the chaos we are causing. It has recovered very well from the disappearance of almost all the dinosaurs. It will recover very well from the coming collapse, and in a million years, new species will have replaced the thousands of species that climate change will have decimated.

It is world peace that is at stake

What is at stake is not the future of the planet, but the future of humanity. Climate disruption will affect our comfort first of all (for most people, heatwaves are tiring, floods are distressing and so on), then our health (with increases in the number of deaths caused by heatwaves and the arrival of new diseases in places that had been free of them). However, very quickly, as we all know, it will be world peace that is at stake.

Box 1.3: One billion climate migrants by 2050?

A World Bank report of March 2018 speaks of 143 million climatic migrants by 2050 and the UN [United Nations] even speaks of 1 billion people moving by this time. These people will be forced to leave their regions because of the direct consequences of climate change (water stress, declining crop yields, lack of food, floods, heatwaves, droughts, cyclones, etc.), but also due to conflicts brought about by these changes.

Some politically unstable states are already experiencing tensions related to climate change, such as the conflict over access to fertile land and drinking water supplies that fuelled the war in Darfur, or the crisis in Syria. Successive droughts that drove rural Syrians to the cities exacerbated social instability and could have contributed to the outbreak of the conflict that led to one of the largest flows of migrants today. The same pattern could be repeated in East Africa.

Without a global treaty on migration, what will be the status of these environmental migrants? Who will help and compensate them?

Pialot, D. (2018) *La Tribune*, 10 July.

Several local conflicts are already linked to problems of drought and access to water. What will happen if there are 200, 300 or 500 million climate refugees? Europe has hosted just over 1 million refugees in the last three years, and the extreme right is back in force in several European Union (EU) member states. What will happen if Africa breaks up and there are tens of millions wanting to cross the Mediterranean to have a chance to live a decent life? Jacques Chirac was right: 'Our home is burning and we are looking elsewhere.'

TWO

Global Warming: The Essential Cause Is Our Greenhouse Gas Emissions

Beating climate change can work only if we identify its causes accurately. This is the second difficulty to overcome if we want to win the race against climate change. After explaining the seriousness of what is coming, we need to clarify the origins of the problem.

A number of citizens (including economic and political decision-makers) are now aware of the gravity of the situation, but they cannot see how to tackle the problem because they think there are doubts about the causes of climate change. On 1 August 2018, a major article published in the *New York Times Magazine* showed that although for the vast majority of scientists, 'the debate on the causes of global warming had already been decided since the early 1980s',[1] very powerful lobbies have done everything to sow doubt, provoke confusion and thus maintain the status quo.

As early as 1979, the Charney Report commissioned by the White House had already announced global warming and stressed the possible impact of human activities on the climate. Nothing in all the new knowledge acquired in the nearly 40 years since contradicts the conclusions of the Charney Report. The IPCC's work only confirmed and refined the analysis, and most of the key points were made in 1979. Unfortunately, with the election of Ronald Reagan to the White House in 1980, the Charney Report was very quickly buried by US leaders. US diplomatic cables made public by WikiLeaks show that the first warning messages date from the early 1970s! As stated in a telegram dated 8 May 1974, referring to the 6th Special Session of the United Nations General Assembly, during which Henry Kissinger took up the problem officially: 'Leaders around the world have expressed concern about indications of possible long-term climate change. Secretary of State Kissinger proposes that this problem be immediately investigated.'[2]

A first warning was issued in Stockholm in 1971: 30 high-level scientists from 14 countries expressed a risk of 'rapid and severe global climate change caused by humans'. In a telegram dated 15 October 1974, the US representative in Geneva reported on the progress of a global atmospheric research programme (GARP), involving a monitoring system (ships, buoys, probes, balloons and satellites), telecommunications and data processing (the first computers). On 1 September 1976, in a message regarding the invitation by US meteorologists to Russian counterparts to participate in working meetings on climate change, 'recently documented proof of global warming' is mentioned.

Two years later, the warning signals were even more convincing, as evidenced by a diplomatic telegram sent to Washington by the US representative at the UN in Geneva on 11 May 1978, probably in view of the first world conference on the climate to be held in Geneva the following year. The diplomat is ringing the alarm bell and worries about leaving 'our children a much more dangerous and riskier world'.

The effects and causes are clearly labelled: 'a warming trend' with consequences that could be 'a disaster for food production'; and an increased risk of 'extreme weather events' that could be due to increased CO_2 caused by the burning of fuels. The term 'greenhouse effect' is launched. In the same telegram, the US representative advocates 'an urgent international program to evaluate the latest knowledge on climate change', the identification of toxic substances polluting the environment and their ban 'if necessary'. It also points to fossil fuels (oil and coal) but believes that the US can still 'depend on them in the short term, maybe a decade or two'.

The US representative states that 'We have seen that they can seriously damage our environment and possibly cause, over the long term, more severe climate change affecting health', and recommends, within a decade or two, 'the safe alternatives that are solar and renewable energies'. The following year, in 1979, the White House commissioned the Charney Report from the Academy of Sciences, which clearly established the link between CO_2 emissions and global warming.

On 24 June 1988, a graph showing the temperature of the globe since 1880 was on the front page of the *New York Times* (see Figure 2.1). The day before, James Hansen, Director of the National Aeronautics and Space Administration (NASA) institute in New York, had been interviewed by Congress, and the scientist's arguments had hit most people by their quality and consistency.

Figure 2.1: Front page of the *New York Times*, 24 June 1988

"All the News That's Fit to Print"

The New York Times

Global Warming Has Begun, Expert Tells Senate

Sharp Cut in Burning of Fossil Fuels Is Urged to Battle Shift in Climate

Global Warming: Greenhouse Effect?

IMMIGRATION LAW IS FAILING TO CUT FLOW FROM MEXICO

ECONOMIC FACTORS CITED

Illegal Entries Are on the Rise as More Come From Large Cities and Stay Longer

Drought Raising Food Prices; Inflation Effect Seems Minor

High Court Getting Unusual Plea Not to Reverse Key Rights Ruling

Source: *New York Times*, 24 June 1988.

From 1978 to 2018 – 40 years lost? In 1978, a senior US diplomat wrote that the US needed to break its dependence on oil and coal 'within a decade or two', but 40 years later, the US President says that climate change is an invention of the Chinese. California is fighting the biggest wildfire ever and burying its dead. How could the Charney Report have been forgotten? Why did Hansen's audition have so little follow-up? Why is the debate still so vague today? Why is action against climate change so weak? The *New York Times Magazine*'s cover story shows how the lobbies of the petroleum industry and conservative lobbies have done everything to make the debate as confusing as possible since the early 1980s. With the Monsanto trial, we have seen how lobbies that want to protect commercial profits are able to act powerfully and in a totally cynical way, ignoring the victims of their actions.[3]

The US is, alas, not an exception. In France too, in the homeland of Descartes and the Enlightenment, we have seen leaders put their intelligence and their networks to the service of the greatest confusion, for example, as in the book published in 2010 by Claude Allègre, the former Minister of Research, who lashed out at climate scientists and said that global warming was 'such an uncertain and useless theory', 'a mumbo-jumbo story'.

Hardly had the book appeared when several hundred scientists reacted against the nonsense and lies in it. The Academy of Sciences was asked to deal with the issue, and a few months later, after a debate open to all who wished to participate, the Academy voted, unanimously, for a text that opposes the thesis of Claude Allègre.

Continuous measurements of CO_2 concentrations began in 1958, when Charles David Keeling, with the help of the US army,[4] installed high-tech instruments at 3,350 metres altitude in an area far from any pollution (city, industrial zone, highway and so on) and began to measure precisely the concentration of CO_2 in the air. The Mauna Loa Observatory is still in use today. Located on the north side of Mauna

Box 2.1: Climate change: the Academy of Sciences refutes the ideas of Claude Allègre unanimously

In a report released on Thursday 28 October, the Academy of Sciences refutes the arguments of Claude Allègre on climate change and reaffirms that the increase of CO2 in the atmosphere is the main cause of global warming.

'Several independent indicators show an increase in global warming from 1975 to 2003. This increase is mainly due to the increase in the concentration of CO2 in the atmosphere,' wrote the Academy in its conclusions, adopted unanimously.

'The increase in CO2 and, to a lesser extent, other greenhouse gases, is undoubtedly due to human activity,' adds the text.

Claude Allègre, member of the Academy, had triggered the controversy with his book *The Climate Imposture*,[5] where he denounced the conclusions of international climatologists. He did, however, endorse this report, which he also adopted.

Le Monde, with AFP (2010) 29 October.

Loa, an extinct volcano in Hawaii, it is managed by the US National Oceanic and Atmospheric Administration (NOAA), a US government agency. It is an essential structure for the measurement of CO_2 emissions in the atmosphere, which has been complemented by other measurement centres distributed throughout the planet. In 60 years, the concentration of CO_2 in the atmosphere has risen from 315 parts per million (ppm) in 1958 to 411 ppm in 2020, 50% more than during the preindustrial period, a considerable increase that can only be explained by emissions from the use of fossil fuels for energy.[6]

On the graph that traces Mauna Loa's records since 1958 (see Figure 2.2), one can see the regular annual oscillations of CO_2 concentration. These are due to the annual rhythm of the

Figure 2.2: The concentration of CO_2 in the atmosphere between 1958 and 2018

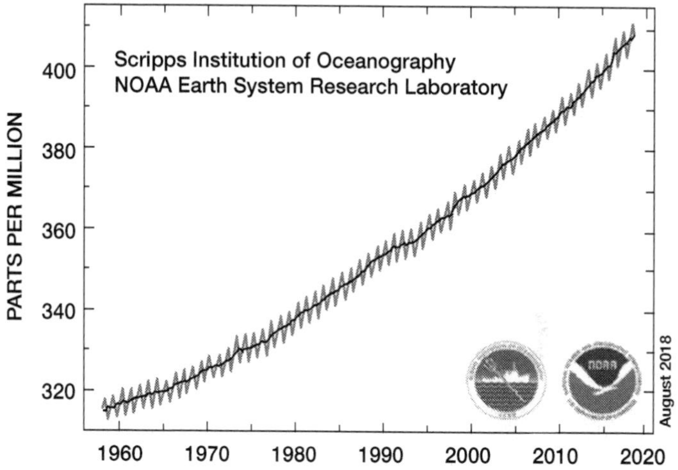

Source: NOAA Earth System Research Laboratory (ESRL) and Scripps Institution of Oceanography.

vegetation. In the spring and summer, the CO_2 concentrations decrease a little because the gas is absorbed by the growing vegetation,[7] as thanks to photosynthesis, plants increase in size by fixing huge amounts of CO_2. On the other hand, during the autumn and winter months, when the leaves fall and start to rot, the CO_2 concentration increases again.

The change in the CO_2 concentration is even more remarkable if we follow it not over 60 years, but over the last thousand years. By analysing the air trapped in the Antarctic ice, the changes in CO_2 can be reconstructed very accurately over a thousand years. The concentration was stable (at around 280 ppm) until the beginning of the industrial era. It then increased faster and faster (see Figure 2.3).

By taking a bigger step back and observing the changes in CO_2 over 800,000 years, one can better understand how our 'development' model is disrupting the balance on our little

Figure 2.3: The concentration of CO_2 in the atmosphere since the year 1000

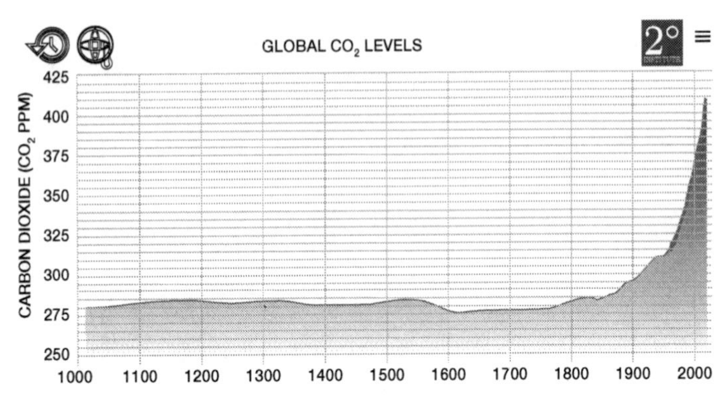

Source: The 2 Degrees Institute.

Figure 2.4: The atmospheric CO_2 concentration over the last 800,000 years

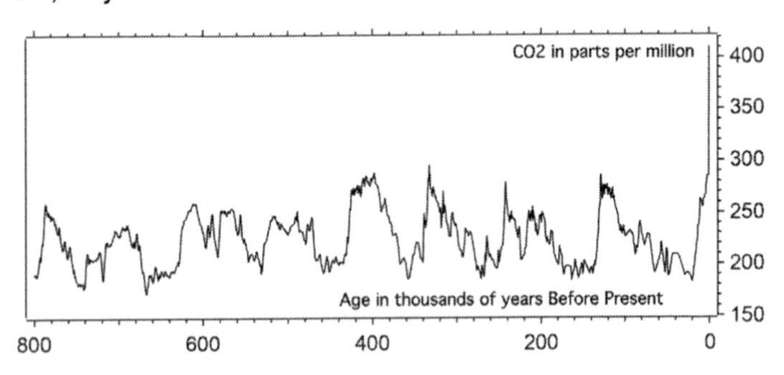

Source: Adapted from Siegenthaler et al (2005) and Luthi et al (2008).

planet (see Figure 2.4). Climate sceptics say that there have always been variations in the concentration of CO_2 in the atmosphere. That is true, but what they 'forget' to say is that not for 800,000 years has the CO_2 concentration exceeded 300 ppm, and that in 100 years, it has gone from 300 ppm to over 400 ppm.

Yes, there have regularly been changes in the temperature of the Earth, especially in relation to its movements with respect to the Sun. However, these movements are much slower and more limited. The last time there was global warming, it was, on average, 20 times slower than the current warming. Moreover, we estimate that our ancestors numbered only 25 or 30 million across the whole planet. They had plenty of time to move, there were plenty of free spaces to welcome them and they had time to evolve. The situation has clearly changed dramatically, with a planet inhabited by more than 7 billion people and global warming being 20 times faster!

The increase in the concentration of greenhouse gases (CO_2, methane and so on) is the main cause of the ongoing warming; there is now almost no doubt of this. As early as 1987, the analysis of Antarctic ice cores carried out by a team of French and Russian researchers, described in a cover story of *Nature*,[8] showed that the increase in CO_2 was not due to natural causes. By analysing the changes in CO_2 over a period of 160,000 years (from air trapped in bubbles in ice for tens of thousands of years), and by reconstructing the temperature at which the ice was formed (which can be determined by analysing the isotopes of hydrogen and oxygen making up the water of the ice), Claude Lorius and his colleagues provided a brilliant illustration of the link between the greenhouse effect and climate that reinforced the theoretical scientific demonstrations on which the Charney Report was based (see Figure 2.5). Looking at the graphs and reading the authors' explanations and the work behind these articles, the link between greenhouse gases and global warming becomes obvious – indeed, indisputable. While most glaciology work published in *Nature* is used by specialists only, all the major media in the world reported on the studies of Claude Lorius and his colleagues, forcing Ronald Reagan, Margaret Thatcher and the rest of the world to accept the creation of the IPCC a few months later.

This historic link between CO_2 and climate, which also applies to methane,[9] over the last 800,000 years,[10] that is, over

Figure 2.5: Variations in CO_2 and methane (CH_4) concentrations in the Antarctic, and of the temperature over the last 800,000 years

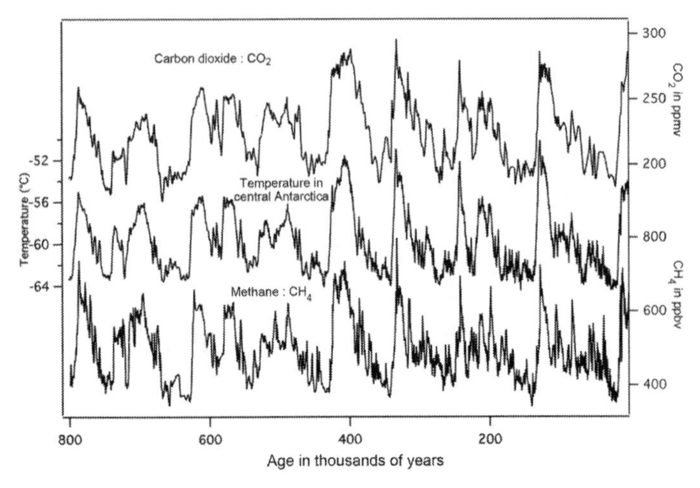

Source: Graphs drawn from the results of the articles cited in notes 8, 9 and 10.

eight climatic cycles, was confirmed. The variations of insolation due to the position of the Earth in its orbit around the Sun are the cause of these glacial–interglacial cycles and determine the pattern, while CO_2 and methane play an amplifying role. This work thus illustrates the importance for the climate of the two main greenhouse gases emitted by our activities over the last two centuries.

To summarize, the increase in the concentrations of greenhouse gases in the atmosphere is the main cause of recent warming; the 'natural' causes unrelated to human activities, such as variations in solar and volcanic activity, cannot account for more than 0.1 °C out of the 0.7 °C observed since the 1950s. In fact, the most likely estimate of the human contribution is that it accounts for all of this recent warming.[11]

At last, some good news:

- First, the causes of the problem are known; the diagnosis is now shared by all serious scientists. Without a proper diagnosis, no action would be effective and all we could do is to weep. Fortunately, this is absolutely not the case: *Eureka!* Thanks to the hard, and not always comfortable, work[12] of hundreds and thousands of scientists around the world, the diagnosis is now certain.
- Second, we are the main cause! Yes, paradoxically, this is very good news, as it is not like the sudden extinction of the dinosaurs. These poor creatures could not do anything to protect themselves since the problem came (it seems) from a huge meteorite that struck the Earth. This cataclysmic collision – a billion times more powerful than the Hiroshima bomb – seems to have caused many volcanos to erupt. In a very short time, the mixture of smoke from volcanoes and dust coming directly from the collision so darkened the sky that the sunlight was blocked, and the plants that fed the dinosaurs died, causing the disappearance of the giants that had resisted the initial heat shock.[13]

This time, the meteorite is us!

The dinosaurs, even if they had had more developed brains, could not have coped with such a fateful blow. Today is different. There is no curse. No foregone conclusion. No meteorite! Or, as some people say with a smile, 'the meteorite is us'. *We* are the cause of the emissions of greenhouse gases. If we decide to stop these emissions or if we reduce them substantially, we should be able to stop climate change. A priori, it is quite possible. However, it is urgent to tackle the problem head-on because, as a growing number of scientists say, 'Soon it will be too late…'.

THREE

'Soon It Will Be Too Late...', Say 15,000 Scientists

> On November 13 2017, more than 15,000 scientists from 184 countries issued a solemn warning against the risks of destabilization of the planet. Bang in the middle of COP23 they raise an alarm call: 'Soon it will be too late….
>
> humanity is not taking the urgent steps needed to safeguard our imperilled biosphere.' Global warming, availability of drinking water, deforestation, reduction in the number of insects and mammals, emissions of greenhouse gases – all the lights are red yet the responses of the political leaders are terribly disappointing.[1]

This text, published on the front page of the paper *Le Monde*, must not be ignored (see Figure 3.1). It sounds an alarm, a wake-up call. Political leaders meeting in Bonn for the 23rd annual summit on climate all know in their hearts that this summit will not change much. As the leaders prepare to 'deal' as usual with reports that have been passed around from government to government for decades, these 15,000 scientists are determined to alert the leaders and citizens of the world of the vital urgency of facing up to the issue and bringing about a radical change of direction. They stress that there are very few years left to change our ways of life if we seriously want to 'save the planet'.

For us, several reasons force us to sound the tocsin and to insist on its very real urgency. There are some who might say that since most scenarios suggest that global warming will not

Figure 3.1: Front page of *Le Monde*, 13 November 2017

Le Monde

Le cri d'alarme de 15 000 scientifiques pour sauver la planète

"IL SERA BIENTÔT TROP TARD..."

Source: *Le Monde*, 13 November 2017.

become really serious for another 20 or 30 years, why not accept a few years of 'business as usual'?

The answer has two parts:

1. Our greenhouse gas emissions continue to increase and the planet seems increasingly unable to absorb them.
2. Feedback mechanisms – very powerful 'vicious circles' – are taking shape, making the impact of these emissions increasingly serious. If we let them continue and they grow more powerful, these 'vicious circles' may soon bring about serious and irreversible changes to life on Earth.

Our emissions continue to increase

When Jean Jouzel and Pierre Larrouturou were auditioned by two National Assembly committees in France, we suggested that the members present should answer a small quiz:

> Knowing that France's objective is carbon neutrality in 2050, which implies a 85 per cent reduction in our

> emissions in 30 years, France should, on average, reduce its CO_2 emissions every year by approximately 3 per cent of their current level. Now, Eurostat has just published a report on the emissions of all the European states in 2017. What do you think France's result is? –3%, –1%, +1%? Go for it. You can vote now.

The parliamentarians looked at each other, smiled and voted. None voted for –3 per cent (clearly, those who thought that all was well did not come to listen). Some raised their hands for –1 per cent and a majority for +1 per cent. However, when we gave the 'right' answer (+3.2 per cent for France and +1.8 per cent for all of Europe), we heard laughter, though for many of the members, it was hollow laughter: the same France that often sees itself as the queen of nations, that hosted COP21 and gave birth to the Paris Agreement, and that organized the One Planet Summit to oppose Donald Trump when he withdrew the US from the Paris Agreement saw its greenhouse gas emissions not decreased by 3 per cent, but *increased* by 3 per cent in 2017!

Some people may believe that we are heading in the right direction – too slowly perhaps, but in the right way. Alas, no, we are on the wrong track! After a few years of slight decline, greenhouse gas emissions from France have started to increase again, as have those from Europe and those from the countries of the planet as a whole (see Figure 3.2). The situation has improved since then with a slight decrease in 2018 and 2019 still largely insufficient with respect to our 2030 objective (by at least a factor of 3). The large decrease (over 10%) observed in 2020 is linked with the COVID-19 pandemic but there is a risk that the recovery plan will be associated with new increases with respect to 2019.

This information should have been front-page news for all the papers and at the heart of political debate. One would have expected the governments would shake up their agendas and organize two-day seminars to draw up action plans adapted to the dangers we are facing, or a European Council specifically devoted to the subject, but no, nothing, *nada!*

Box 3.1: Worldwide CO_2 emissions take off again

Behind the promises of decarbonisation of the economy, greenwashing and political speeches lies the inexorable reality. After three years of stagnation, in 2017 emissions of carbon dioxide (CO2) are rising again, dashing hopes of seeing humanity starting to decrease its emissions. This is the major finding of the work published on 13 November by the Global Carbon Project (GCP) consortium, that reports on CO2 emissions annually.

According to the GCP, this year should end with a total of about 41 billion tonnes of CO2 (gigatonnes, or GtCO2) emitted by the burning of fossil fuels, industrial activities and land use – essentially deforestation. An increase of 2% in one year.

China is still the prime emitter: 10.2 GtCO2, a quarter of global emissions. Next comes the United States (5.3 GtCO2), India (2.4 GtCO2), Russia (1.6 GtCO2), Japan (1.2 GtCO2) and Germany (0.8 GtCO2).... France arrives at 19th place. Taken as a whole, the European Union comes in third, with emissions of 3.5 billion tonnes of CO2.

Roger, S. and Foucart, S. (2017) *Le Monde*, 13 November.

However, there is worse news. Going back to the graph tracing the CO_2 concentration in the atmosphere since 1958 (see Figure 2.2), we can see that the time required to cross a '10 ppm marker' is getting shorter and shorter:

- from 320 ppm to 330 ppm took twelve years;
- from 330 ppm to 340 ppm took eight years;
- from 340 ppm to 350 ppm took six years;
- from 350 ppm to 360 ppm took seven years;
- from 360 ppm to 370 ppm took six years;
- from 370 ppm to 380 ppm took five years;
- from 380 ppm to 390 ppm took five years;
- from 390 ppm to 400 ppm took five years; and
- from 400 ppm to 410 ppm took only four years!

Figure 3.2: CO_2 emissions in gigatonnes per year

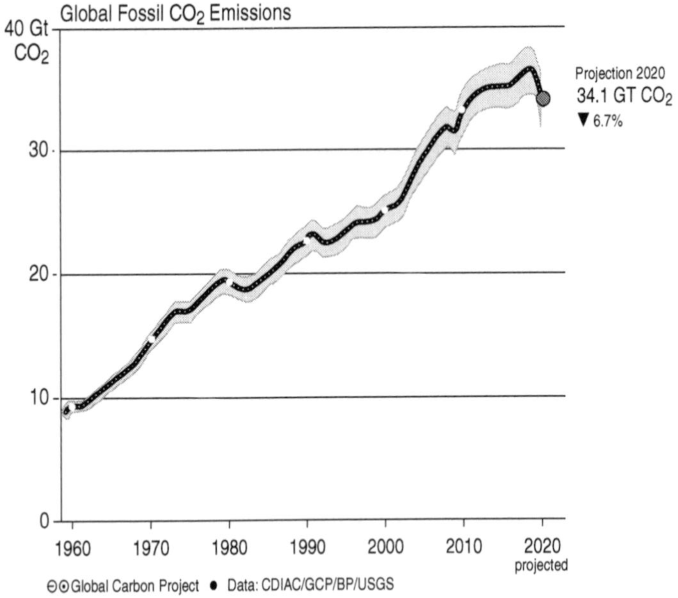

Global Fossil CO_2 Emissions
40 Gt CO_2
30
20
10
0
1960 1970 1980 1990 2000 2010 2020 projected
Projection 2020
34.1 GT CO_2
▼ 6.7%
Global Carbon Project • Data: CDIAC/GCP/BP/USGS

Note: Global CO_2 emissions have increased considerably since 1960. An improvement seemed to be on the horizon: greenhouse gas emissions from the burning of fossil fuels (coal, oil and gas) and from the cement industry, that is, 70 per cent of total emissions, had stabilized from 2014 to 2016, but they then increased again in 2017, 2018 and 2019. The decrease observed in 2020 is linked with the COVID-19 pandemic.

Source: Global Carbon Project, 2020.

Despite all the international summits, the COPs and the Paris Agreement, and despite the number of extreme weather events, which have more than doubled – sharp warnings that should impel us to change our model – not only does the concentration of CO_2 in the atmosphere not go down, and not only does it not stabilize, but *its rate of increase is three times faster than it was 60 years ago!*

For a long time, the planet behaved well, or was a good mother: she absorbed most of our CO_2 emissions in the oceans and in the forests. However, this capacity to absorb

Box 3.2: Deforestation: the anatomy of an approaching disaster

Every year, 13 million hectares of forests disappear around the world, the equivalent of a quarter of France. Deforestation is responsible for 20 to 25% of global greenhouse gas emissions. To be specific, the destruction of tropical forests releases more to the atmosphere than the entire transport sector worldwide. In the countries of the South – where since 1990 forests have been cleared more than in the North – 35% of greenhouse gas emissions are due to deforestation. This rises to 65% in the poorest regions.

Agriculture is the major cause of this phenomenon. It is responsible for 80% of deforestation in the tropical zones.[2] Agricultural production can take different forms: livestock, cultivation of soya or palm oil, rubber, or the production of pulp for paper.

Blavignat, Y. (2017) *Le Figaro*, 19 July.

CO_2 is decreasing visibly. It is not surprising, as we have never destroyed so many forests.

If the 15,000 scientists are ringing the alarm bell, it is not just because the concentration of CO_2 in the atmosphere is increasing much too fast, but also because they fear that we will soon be facing chain reactions ('domino effects') that will accelerate climate change, which may then become uncontrollable.

The mechanisms behind what researchers call 'feedback loops', or 'vicious circles', are quite easy to understand. Take the example of sea ice in the Arctic. We all know that a white surface reflects almost all the light and energy it receives, while a dark surface absorbs most of the energy received. It is easy to understand that when ice and snow melt, white surfaces are progressively replaced by dark surfaces that absorb much more heat.

Now, the melting of the Earth's ice is accelerating (see Figure 3.3). This amplifies warming at the higher latitudes of the northern hemisphere, and the Arctic is warming twice

Figure 3.3: Changes in the area covered by ice in the Arctic Ocean and in the Southern Ocean between 1979 and 2017

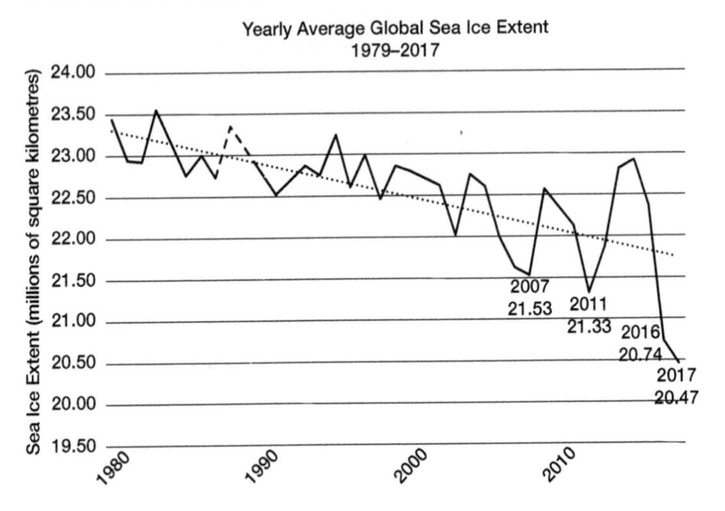

Source: Data from the NSIDC Sea Ice Index V 3.0 (available at: www.weathernetwork.com). Data missing from December 1987 and January 1988.

as fast as the Earth's average. This means that the melting of the ice is accelerating in Greenland too, and this mass of ice is contributing to the rise in sea levels.

The Antarctic ice seemed to be resistant to global warming. Alas, this is no longer the case.

In less than 40 years, the total area of sea ice, that is, both Antarctic and Arctic combined, has gone from nearly 23.5 million km^2 to less than 21 million km^2. This change is striking: it is as though an area four times the size of France has changed colour and stopped radiating back the heat it receives from the sun. Global warming melts ice, and melting ice reduces white surfaces, which, in turn, accelerates global warming, and so it goes on (see Figure 3.4).

There is another vicious circle that keeps many scientists awake: the melting of the permafrost. Permafrost is the frozen earth existing around the Arctic Circle in Northern

Box 3.3: Antarctica: the melting of the ice has accelerated over the last five years

Antarctica has lost 3,000 billion tonnes of ice since 1992, and this trend has accelerated dramatically over the past five years, according to a study published in *Nature* on Wednesday.

With more than 98% of the continent covered by permanent ice, Antarctica alone accounts for 90% of terrestrial ice. If all this ice melted, it would raise the level of the oceans by almost 60 meters.

Until now, scientists have been at pains to determine whether Antarctica had gained mass as a result of snowfall or lost it due to melting ice or the dislocation of icebergs.

But satellite observations over two decades have provided a more comprehensive view. Prior to 2012 Antarctica, at the South Pole, lost about 76 billion tonnes of ice annually, according to the baseline study by 84 scientists. Since then, that figure has leaped to 219 billion tonnes a year. In other words, for the past five years, the ice has melted at almost three times the previous rate.

This discovery should dispel any doubts that Antarctica is melting rapidly and poses a threat to hundreds of millions of people living in low-lying coastal areas. 'It should be a very great cause for alarm', says Martin Siegert, professor at Imperial College London, who did not participate in the study. Now it is up to policy makers to act.

Ouest-France (2018) 13 June.

Europe, Russia, Northern Asia and America. Permafrost has been frozen for thousands of years, and the organic matter it contains is protected by the cold, like food in a huge freezer. However, with the current warming, thousands of hectares of permafrost are defrosting and its organic matter is thawing and decomposing, producing large amounts of CO_2 and CH_4 (methane, a gas with 30 times the heating capacity of CO_2). It is estimated that permafrost melt could emit 1.5 billion tonnes

Figure 3.4: The melting of the ice caused by global warming accelerates this warming

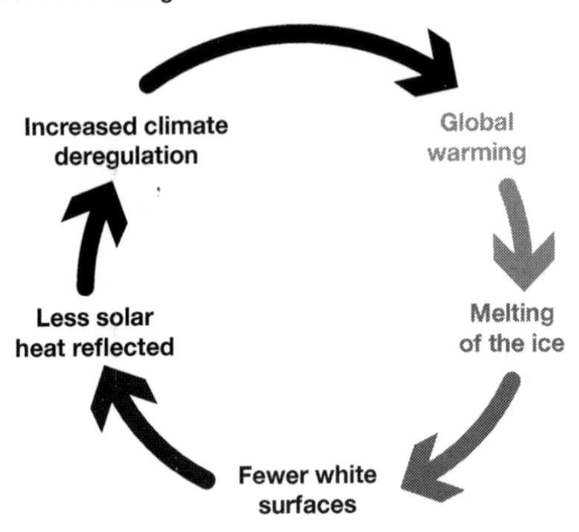

of CO_2-equivalent greenhouse gases each year, that is, three times the emissions of France.

A third example of 'feedback' that should impel us to act without delay is the increase of forest fires already related to global warming. The extent of the problem is clear from the statistics of forest fires in California, where the areas consumed by fires have been measured annually since the 1930s (see Figure 3.5).

The graph in Figure 3.5 shows that burned areas stayed roughly constant during the 1940s–1970s and began to increase slightly in the 1980s. However, in the last 20 years, the increase has been spectacular, and given the severity of the fires in the summer of 2018, one fears that the forecasts for the entire 2010–19 decade, published at the end of 2017, have sadly been overtaken by reality.

'California's five years of drought have left more than 100 million trees dead', said a Berkeley scientist:

> This is obviously not the only issue, maybe the forests need to be managed differently, but when there are so many dead trees, fires are very difficult to fight! What's more, with global warming, summers are longer and

Figure 3.5: The number of hectares burned by forest fires in California since the 1940s

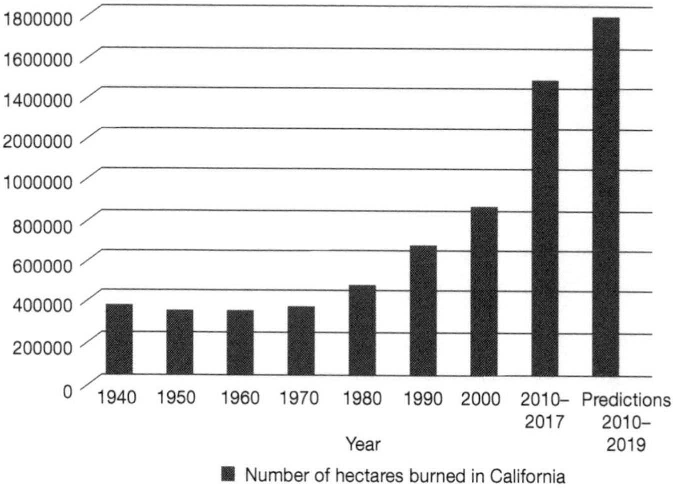

1800000
1600000
1400000
2000000
1000000
800000
600000
400000
200000
0
1940 1950 1960 1970 1980 1990 2000 2010–2017 Predictions 2010–2019
Year
Number of hectares burned in California

> warmer. Previously we had fires mainly in September and October. Now, the fires begin as early as July.[3]

Whatever happens at the end of 2018 and in 2019, the area burned will have been multiplied by four or by four and a half in the space of 50 years, which is considerable. We all know that a growing forest absorbs tonnes of CO_2 and that an 'adult' forest stocks a very large volume of CO_2, but a burning forest releases a huge amount of CO_2 into the atmosphere, and this then becomes a black surface that absorbs heat.

Global warming is one of the causes of the spread of forest fires, which have become increasingly violent and deadly in California, as in many other parts of the world. These forest fires release a considerable amount of CO_2, which increases global warming, and so it goes on.

'Soon it will be too late...' – the 15,000 scientists are right to ring alarm bells. By continuing to emit more and more greenhouse gases and accelerating the deforestation of our fragile planet with very powerful 'vicious circles' already active, we will soon reach a point of no return, a point of rupture beyond which our efforts would be in vain. It would no longer be possible to avoid collapse.[4] 'Like sleepwalkers, we are heading for the abyss.'

FOUR

The UN Environment Programme Denounces 'This Catastrophic Climate Gap' between the Reductions Needed and the National Pledges

In November 2017, the UN Environment Programme issued a warning of 'this catastrophic … gap' between the current pledges of the states to reduce their greenhouse gas emissions and the efforts needed to comply with the Paris Agreement – to contain global warming well below 2 °C and to endeavour to limit it to 1.5 °C.[1] In general, UN diplomats communicate in a more consensual manner, but this time, they were seriously worried and did not hesitate to say so. Major media around the world have echoed the following statement by the Director of the UN Environment Programme: 'governments, private sector and civil society must bridge this catastrophic climate gap'.[2]

Box 4.1: Climate: the battle for 2°C is practically lost

The battle for the climate is not yet lost, but it has started very badly.... This is not the first warning given by UN Environment but it has a particularly urgent tone, just a few days before the opening of COP23 and after a cataclysmic summer, during which a succession of hurricanes, floods and fires showed the vulnerability of both rich and poor countries to climate disruption.

The commitments made in 2015 by the 195 countries involved in the Paris Agreement will only make it possible to do 'About a third' of the job, warn the authors.

Assuming that all the states respect all their promises, which are non-binding and sometimes conditional on obtaining international funding, the Earth is now moving towards a rise in the thermometer from 3°C to 3.2°C at the end of the century.

UN Environment still tries to remain optimistic. In their view it is still 'possible' to avoid widespread overheating. 'The report shows how a shift in technology and investment can reduce emissions, while creating huge social, economic and environmental opportunities', says Norwegian director Erik Solheim.

Humanity has not yet burned all its bridges. But we have entered the danger zone.

Le Hir, P. (2017) *Le Monde*, 31 October.

Figure 4.1 shows the four scenarios studied by the UN Environment Programme:

- The first, most pessimistic, scenario is a complete lack of response by human societies: greenhouse gas emissions continue at the same rate as for the last 20 or 30 years. A 1.5 °C increase would be reached by 2040 and we would suffer a temperature rise of 4–5 °C by the end of the century. The Earth would become an oven. *No comment!*
- The second scenario is hardly less sinister since it leads to a warming of 3–3.5 °C, with a dramatic impact on water cycles and harvests. Alas, this second scenario assumes that 'States respect the commitments they have made to implement the Paris Agreement'. We already know that we are not honouring these commitments at the moment, and the UN reminds us that even if we were respecting them,

their impact would be totally insufficient to prevent today's young people from facing a warming in the second half of this century that would be very difficult for them to adapt to – if not impossible in some parts of the globe.

- In the third and fourth scenarios, to contain global warming below 2 °C, global greenhouse gas emissions should be capped at 42 billion tonnes in 2030, compared with a total of about 52 billion tonnes of CO_2 and methane emissions in 2016. A maximum of 36 billion tonnes should be the 2030 target for there to be any hope of staying below 1.5 °C.

Figure 4.1: Predicted rise in temperatures as a function of different trajectories of our emissions

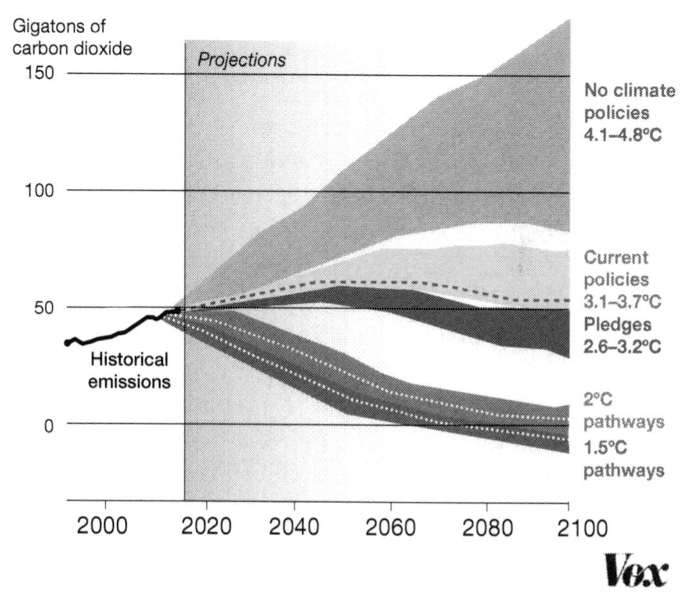

Source: Climate Action Tracker.

Recent studies – which the UN Environment Programme said at the end of 2017 will be used in future reports – conclude that a much lower level, about 24 billion tonnes in 2030, will be needed to avoid going over 1.5 °C. The most radical solution is known, the UN Environment Programme reminds us in its report: this requires leaving underground 80–90 per cent of the coal reserves, half of the gas and about one-third of the oil reserves. This means no longer building new coal-fired power plants and programming the shutdown of nearly 6,700 units currently in service. However, say the authors, 'other levers must also be activated. Just by promoting solar and wind power, improving energy efficiency, developing alternative modes of transportation, stopping deforestation and doing reforestation could reduce annual emissions by 22 billion tonnes.'

Reading this study, we understand that the Paris Agreement was, without question, very successful diplomatically: all the leaders of the planet signing a text with the goal of limiting warming to 1.5 °C was a real advance. However, it is a bit like the Maginot Line – almost all our leaders thought in 1939 that the Line would protect France very effectively from the Nazi army. Implementation of the Paris Agreement, sticking to these commitments, will not solve the problem.

The measures that should be taken are analysed in the IPCC Special Report on 'Limiting to 1.5°C in the context of sustainable development, poverty eradication and equity'; its executive summary for policymakers was approved and published in early October 2018. This special report demonstrates the clear benefits of staying below 1.5 °C rather than 2 °C in terms of extreme events, rising sea levels, loss of biodiversity, agricultural yields, access to water, human health and so on.

But what a challenge! To have better odds than two out of three to achieve this, we have about a dozen years left with CO_2 emissions at the current levels, provided that we are then able to move to zero emissions immediately. This is obviously impossible, so we need an extremely rapid reduction of our CO_2 emissions now and zero emissions by 2050. At the same time, we must also achieve significant reductions in other greenhouse gases, especially methane.

For many of the emission scenarios analysed, meeting the 1.5 °C target implies the implementation of negative emissions; for example, by combining the use of biomass for energy with the trapping and storage of CO_2. This, of course, implies competition for land – for producing energy or for food.

If we want to avoid chaos, everything remains to be done, or almost everything. As far as France is concerned, it must change its development model radically in order to respect its commitments – at European level – to reduce emissions by 40 per cent[3] – or better by 50 per cent – by 2030 (this is tomorrow) and reach zero net emissions in 2050. Moreover, these goals must be shared by the whole of Europe.

There is no more time to lose. We must stop making empty speeches, stop trying to patch up the existing model and declare war on climate disruption.

FIVE

Zero Net Carbon Emissions? Yes, It Is Possible

As early as 2003, France had set itself a target to reduce its carbon emissions by a factor of four.[1] In July 2017, Nicolas Hulot set an even more ambitious goal of 'carbon neutrality'. To attain carbon neutrality within 30 years, radical changes are needed: all economic sectors must halve their energy consumption and what remains must be 'carbon-free' energy. Zero net carbon emissions means the balance of emissions is nil, but this does not mean that no carbon is emitted – that would be utopian. The objective is to reduce our emissions drastically and for the remaining emissions of certain sectors to be absorbed by other sectors; in this way, instead of being released into the atmosphere, methane could be 'captured' and used to produce fuel for clean transport.

To achieve this revolution in carbon emissions, the solution can be summarized in three points. We need to:

1. save energy
2. save energy
3. replace fossil fuels with carbon-free energy sources.

Sufficiency and efficiency

Sufficiency and efficiency are the two keywords of the 'negawatt' scenario. We might expect that a think tank composed of many experts, engineers, technicians and researchers would extol the merits of a specific technology, but that is not the case. No technical solution will allow us to attain zero net emissions if we are not capable of massively saving energy, thanks to sufficiency and efficiency.

We need to change our behaviours and be frugal. "If you don't have a Rolex by the time you're 50, your life is a failure", said a famous French publicist. '*I spend therefore I am*' – is this humankind's future? Is it not urgent to change our field of imaginary, freeing ourselves from the 'more is better' mantra, of Donald Trump's dream of a society where everything is '*great*', excessive and shiny, and of a society of possessions that urges everyone to purchase more without really living better?

Give up the superfluous? To address the climate challenge, we must be capable of a certain restraint in our behaviours, our purchases and our investments:

> 'The first Fiat 500 weighed 450 kg, now a Fiat 500 weighs about 900 kg, as a member of the French Physics Society who studies such matters pointed out. Twice as much! Is it really necessary? And the Fiat isn't an exception. In total, millions of extra tonnes of steel must be produced each year to build our cars, that then need to be propelled on roads.... Can we have cars that are considerably lighter, even if they don't go as fast?'[2]

"Personally, I don't really like the word sobriety", says one physicist colleague: "I'm afraid that it will be misunderstood: it seems ideological, religious.... But in substance, I entirely agree: if we don't want authoritarian or violent restrictions (cars will be totally banned or too expensive and, as such, reserved to a limited few), we need to willingly reduce our consumption." Restraint is both a series of personal choices (eating less meat, flying less or only exceptionally, continuing to wear a shirt when its colours are a little washed out, recycling paper, participating in the neighbourhood's public composting initiative and so on) and a series of collective initiatives to step away from a society whose ecological footprint has become unacceptable.

Repairing, recycling and reusing

Aiming for greater restraint consists in, for example, developing circular economies, as some communities, organizations and businesses have already done on a small scale. When a washing machine breaks down, we generally hear: 'It will cost you more to repair it than to buy a new one.' If we are conscious of living on a finite planet, can we keep on thinking this way? Everywhere in France, the employees of Envie recover electrical appliances that people are tempted to throw away. They repair and resell them, and when the oven or the washing machine cannot be repaired, they recover what can be salvaged: copper, electrical circuits, steel, plastic and so on. This business employs hundreds of individuals in social reintegration. Instead of wasting limited resources by throwing away what should not be and buying a new 'Made in China' machine – made and transported in dubious social and ecological conditions – we can create local jobs and contribute to a better use of scarce resources.

In all regions, within the networks of the social and solidarity-based economy, similar initiatives are happening. Why do we not accelerate this momentum? There are tens of thousands of jobs waiting to be created. Useful jobs that cannot be outsourced.

We can go even further and develop a service-based economy in certain sectors. Such an economy, based on serviceability, sells the *use* of a good and not the good itself. The producer of the good is then inclined to make their product last as long as they can instead of programming its rapid obsolescence in order to sell another one in a few years.

The Xerox company, for example, decided a few years ago to no longer sell photocopiers, but to make them accessible to their clients and to bill their clients for their use. By remaining the owners of the machines, Xerox obviously benefits from their prolonged durability and has reviewed the manufacture of the machines by applying the 'dismantlable, reparable and reusable' principle. Today, thanks to this principle, their new generation of photocopiers are composed of 70–90 per cent of parts of old machines. This gives rise to significant environmental and financial gains.

Why not apply this principle to the automobile market? Instead of buying a car, consumers would buy the right to use the car for a whole year, or whenever they need it. Hence, the numbers of vehicles built each year and the costs (including insurance costs), as well as the direct and indirect pollution, would be reduced and shared better.

To decrease our need for energy, steel and rare earth minerals, we can develop systems for car sharing in certain geographical areas: some people, no doubt, need a car practically every day, but others only need one exceptionally. Buying a car and insuring it for the entire year is an excessive cost considering the limited number of days where it is used. Furthermore, it would be more useful and convenient to be able to choose between a small car to transport one or two people, or a larger car to transport more people or bigger objects, which should lead to favouring self-service car-rental systems. 'It's impossible!', some will say, 'You're daydreaming.' However, they said the same things 30 years ago when we were talking about putting bike-sharing systems in Toulouse, Lyon and Paris.

Halving our greenhouse gas emissions by 2030? Reaching net zero emissions by 2050? It is a colossal task that should not scare us; indeed, in each of our countries, we know the origins of CO_2 and of methane released into the atmosphere, and we can (and must) systematically address each of these sources.

In France, to attain the objective of 'zero emissions', the law on energy transition states that our energy consumption must be halved by 2050. This same law has also set benchmarks to be attained by 2030, as well as the objective of reducing by 50 per cent the share of nuclear energy in electricity production by 2025. These objectives concerning nuclear energy and the halving of our energy consumption by 2050 have led to intense debates. Reducing the share of nuclear energy will not be easy – this has been illustrated by the abandonment of the 2025 target for reducing the share of nuclear energy in electricity production by 50 per cent – but it will favour the development of renewable energies that are destined to have an increasingly important worldwide impact.

Simple measures, at least initially, like immediately closing the remaining coal plants that are still working, can be put into effect rapidly. However, to halve our energy consumption, radical proposals must be considered. The negawatt scenario reminds us that the best energy is that which is not used and pushes the development of renewable energies.

Europe is also ambitious in announcing the 2050 objective of dividing by five its emissions compared to the level in 1990. However, it seems vain to hope that this goal will be achieved without a real European energy policy such as the one we are pushing for. At the global scale, in most scenarios that would keep global warming under 2 °C, the share of low-carbon electricity must increase from 30 per cent today to over 80 per cent by 2050.[3]

Housing, transport and agriculture: a full review of our model

In France, the main source of greenhouse gas is buildings, that is, the houses in which we live and the offices and factories in which we work. The efforts made to insulate these buildings remain tragically insufficient, even though the cost-effectiveness (ecologically as well as economically) of a high level of insulation has been demonstrated. Street by street, in all our neighbourhoods, in all our cities and towns, we will need to insulate our houses, our offices, our factories and all the stores, insulate the gyms, the theatres, the churches, the farmhouses, the town halls, the schools and universities, and so on. Transport is also an important source of greenhouse gas, comprising 26 per cent of our emissions. We use our cars too often, including for short trips that could be made by walking or biking. Public transport is still insufficiently developed or unreliable.[4] We eat food that has sometimes travelled thousands of kilometres: 'locally produced, seasonal food' represents an insufficient share of our diet.

Thousands of trucks are travelling along our roads and highways, emitting a colossal quantity of CO_2 (and fine particles that damage our lungs and those of our children). There would be fewer trucks if we actively developed short supply chains and if we invested massively in rail freight, as some countries like Switzerland have successfully done.

Agriculture also produces a significant share of emissions. This would not be the case if we decided to change our eating habits progressively. Emissions can also be reduced if the global agricultural system reduced its use of inputs (fertilizers, pesticides and so on) and valued farm residues by producing renewable energy in the form of methane to be used in transport and housing.[5]

To start thinking about the evolution of livestock farming, we can find inspiration in what happened in wine growing: in a few decades, wine consumption per capita has reduced by more than two-thirds, dropping from 140 litres per person per year to 40 litres today. Has it been detrimental to winemakers? No. However, they have abandoned mass production of cheap wine and have shifted towards smaller quantities of better quality wine. Conversely, if we want to fight deforestation in countries of the South and against greenhouse gas emissions in the world's countryside, why not produce less meat of better quality.

If the goal is to achieve zero net emissions, we need to insulate 100 per cent of our housing stock: insulate all our buildings, public and private, except those that need complete reconstruction … in wood.[6] We must reassess 100 per cent of the food chain and rethink 100 per cent of our transport and 100 per cent of our leisure activities.[7] Obviously, none of us has the skills to address the numerous questions that will arise. However, in each sector, both in France and in many other countries, pioneers have been talking about these problems for many years now, and we already have many solutions that need increasing in scale.

Sobriety and improved efficiency should allow us to halve our energy consumption. However, decarbonizing the energy that will be used in the future remains to be achieved.

Investing massively in renewables

This is where the massive development of renewable energies comes in, allowing us to keep most of the remaining oil, gas and coal underground. What makes us optimistic here is the incredible progress that has been made in a few years to decrease the cost of renewable energies: the price of a kilowatt-hour of electricity coming from solar panels has been reduced tenfold in ten years – that is a real revolution.

However, some grouches will object that the problem with renewable energies is their intermittence:

> It is all very good, solar and wind. It is true that we have made colossal progress in less than ten years, but a nuclear powerplant produces electricity 24/7. Solar and wind are not continuous, nor can they be adjusted to demand: there are periods when they produce excessively and we do not know what to do with the surplus, and there are times where the wind does not blow or the sun does not shine. What should we do in these periods?

Our friends the grouches are right and highlight one of the well-known limits of renewable energies. What can we say? We offer three points:

- If we significantly diminish our consumption (through sobriety and efficiency), consumption peaks, which require starting the most polluting thermal power plants, should disappear. If, in ten years, we have thermally insulated all the buildings that leak heat,[8] people in charge of the electrical grids can deal with the 6pm to 9pm time slot (where everyone comes home from work and turns on the heat in the wintertime, while turning on the TV and the oven, and so on) without needing thermal power plants.

- There is not just wind and solar. If we develop all renewable energies without hesitation, including hot-water solar (which has a very efficient yield), heat pumps, geothermal and biomass, we have quite a balanced mix of energies and we are less dependent upon sunlight and wind. Contrary to popular belief, not all renewable energies are intermittent, for example, biomass and geothermal energy are not produced irregularly.
- We have made colossal progress in (renewable) energy storage. In only five years, the cost of storing energy in a battery has been reduced almost threefold – in only five years!

"Decrease in the production and storage costs.... We see that a shift is happening", explains a renewable energy expert: "In the French overseas territories, islands where electricity had always been produced by thermal power plants using fossil fuels, there is a shift towards renewable which is cheaper, even when taking into account the costs of installation and storage." Furthermore, batteries are not the only way to store energy that is occasionally produced in excess by wind and solar sources. When Denmark produces too much electricity because the wind is blowing very strongly, they send the electricity to Norway. Norway uses it to bring water back upstream behind its dams by running a turbine backwards. The water that is stored upstream will then be used to make electricity when the wind is insufficient.

In another way to store energy, excess electricity can be used to produce hydrogen by classical hydrolysis, as is taught in high school. This hydrogen can be used to produce steel (which avoids the use of coal in the reaction that transforms cast iron into steel), or for heating by adding small quantities to mains gas, or to fuel urban transport with a fleet of buses or trucks that run on hydrogen. However, the hydrogen produced when wind and solar produce 'too much' electricity can also be used to produce methane through a reaction that uses CO_2 in the air, and this methane will itself serve, directly or after transformation into methanol, to propel hybrid cars (renewable electricity and renewable gas).

Anyone who has had the chance to spend two or three hours with researchers who are working on these matters understands that we are living through a real revolution, and that we are, probably, only at the beginning! A tipping point is approaching. There have never been so many researchers involved in this subject area. Whether Donald Trump likes it or not, even in the US, the number of patents for renewable energies exceeds those for fossil fuels. Yesterday's problems are tomorrow's solutions.

Investing in research

There remains one important research effort that must be accomplished. No one can avoid the difficulties that must be overcome, but it is probable that we are very close to living through a revolution in energy terms that is comparable to the internet revolution: at first, there are a few hundred or a few thousand researchers who seem a little cut off from normal life, a few *geeks* passionate about their subject; then, in a little less than 20 years, there are billions of men, women and children using what has been conceived in labs by these unknown researchers. No one can deny the transformative power of the internet revolution, which impacts almost all aspects of life. However, it is probable that such a revolution is happening in the energy sector, and if Europe decided to give itself the financial means to win the war against climate change, it could become a leader in the research and industrial sectors on these matters. Michel Spiro was right: on top of what already exists, if we want to win the race against climate change, it is urgent to create a real community of researchers dedicated to this battle, as we did when the European Council for Nuclear Research (CERN) was created for research into matter's fundamental particles.

After Airbus and Ariane, the car of the future?

On some subjects, most solutions are already known: we need only to act with budgets that will allow us to change gear quickly and radically. However, in other areas, we need to invest radically in research. Europe invented Airbus and Ariane. With CERN, Europe has invented the internet and touchscreen computers. Why should Europe not be capable of inventing the car of the future, a computer that uses significantly less energy and so on? Why would it not be capable of inventing the car of the 2030s: a lighter car and, probably, a car that will be a little slower but use much less energy, 100 per cent of which will be renewable?

We obviously should prioritize walking and biking (it is good for the planet and far better for our health[9]), and we should massively develop public transport, but we do need personal mobility occasionally or regularly. For this personal mobility to be compatible with the zero emissions objective, it is urgent to give research the necessary means to find the solutions. Without public investment, Ariane and Airbus would never have seen the light of day. Why is Europe slow in investing in tomorrow's transport systems?

Yes, it is a colossal task that we have ahead of us: changing a million little things in our daily lives and, in parallel, launching gigantic collective projects. To face a threat whose scope has never been seen before, we need to launch a project of adequate scope.

SIX

Can We Make a Colossal Development Programme Work? We Can Do It!

To avoid climate chaos, we must succeed in a huge programme of development. This should not scare us because, in recent times, people have already been able to carry out colossal programmes at this scale: like 'education for all', achieved in France's Third Republic. When France decided to make education accessible to all, the country was able to make such a huge effort: in all towns and villages of more than 500 inhabitants, a site was found, if possible, in the centre of the town, and a boys' school, then a girls' school and then the teachers' houses were built. Here, we are far from the theoretical debates in Parliament; there is only the smell of mortar and sweat. Sometimes, tempers flared because they met unexpected difficulties; for example, falling behind the planned timing or the mayor disagreeing with the architect, the builder or the carpenter. Here and there, things did not work: people got tired and there was a shortage of workers. However, overall, they won: what seemed impossible to some, both technically and financially, became a reality. In the 19th century, the French Republic built tens of thousands of schools (*tens of thousands!*), and the tens of thousands of teachers needed for the success of the project were trained, paid and housed.

In 40 years, in a country much less affluent than it is today, the number of schoolchildren more than quadrupled. School for all? An impossible task? *Impossible is not French!* Moreover, some countries have been able to move even faster on massive development programmes than France in the Third Republic, for example, like the US between 1941 and 1945.

Victory Program versus America First

On 7 December 1941, the Japanese air force bombed the US fleet at Pearl Harbor. The toll was dramatic: 2,403 dead and more than 1,000 wounded. The next day, President Franklin Delano Roosevelt asked Congress to declare a state of war against Japan. On 20 December, the mobilization extended to all Americans between 20 and 44 years: 10 million conscripts and 5 million volunteers then joined the US army. These GIs were joined by more than 250,000 women. To arm them, the entirety of US industry was mobilized; it was the beginning of an incredible war effort *never seen before!* To avenge Pearl Harbor, the American people supported Roosevelt immediately – until that date, neither the American public nor the army were ready to go to war.

An isolationist committee, America First, was formed in 1940 and campaigned actively for the US to stay out of the conflict. However, after Pearl Harbor, America First was in trouble. The next day, President Roosevelt decided to put the entire US production system at the service of the war effort: to produce tens of thousands of tanks and fighter jets, he forced the car industry to stop producing cars in order to focus exclusively on the production of war machines. The bosses and the shareholders of the automobile industry were not immediately convinced, but Roosevelt left them no choice. By putting the most efficient part of the economy at the service of the political objectives to which a large majority of citizens agreed, Roosevelt succeeded, and the production of aeroplanes and tanks exceeded the initial objectives. In three years, the factories produced 275,000 aircraft, more than 6 million light vehicles, 90,000 tanks and 65 million tonnes of shipping.

In 1942, Roosevelt created the War Production Board (WPB) to regulate and distribute the production of materials and fuel until the end of the war. To manage the Victory Program, the number of civil servants was increased from 1 million to 4 million. The first problem that the WPB had to solve was the shortage of raw materials because the US was cut off from their sources of supplies in the Far East, in particular, South East Asian rubber, essential for war vehicles. The WPB imposed restrictions on production for civil use: in order to save copper, the production of household appliances was practically stopped. Those whose washing machines no longer worked had to wash their clothes at their neighbour's house.

In just a few months, the *conversion* was achieved, even the main pencil factories started producing bombs. The production of cars was stopped because factories were needed to manufacture aircraft engines. From the large car factories emerged not a single car, but rather tanks or planes. To save rubber, the WPB broadcast appeals encouraging citizens to take public transport or to 'carpool between neighbours' rather than driving their own cars in order to avoid tyre wear. Carpooling? In 1943? Without the internet or BlaBlaCar? Yes, it was possible! The WPB also launched a major campaign of rubber recovery and organized the large-scale production of synthetic rubber.

President Roosevelt relied on the strongest part of the economy – the automobile industry – and he 'twisted' it, changing its orientation to put it at the service of the common good; in these years of war, this was the fight against the armies of the Axis Powers. In total, the number of employees increased by 30 per cent (and 65 per cent in factories!), while 12 million men left the labour market to join the armed forces. To solve the problem of the shortage of labour, many new 'types' were hired, including black people and women.[1] Some said: 'What about training?' Many of these women and black people had never worked in factories. How could one hope to accelerate the pace of production by hiring people without the right training? Here too came innovation: since there was no choice, employees were trained on the job, 'hands on'. No one blamed them for mistakes, and their productivity very quickly caught up with that of the former employees. Under these conditions, unemployment dropped massively and wages increased.

The huge programme launched by Roosevelt was a double success:

- it allowed the Allies to defeat the German, Italian and Japanese armies; and
- it allowed the US to regain full employment, contrary to the arguments of America First, who claimed that entry into the war and solidarity with the French, the British and the Chinese would ruin the US.

To win the war against climate change, we need to roll up our sleeves and launch a programme as gigantic and as successful as the US Victory Program.

'We choose to go to the Moon'

A third example of a programme of a type that had *never been seen,* and that worked successfully, was the space programme launched by John F. Kennedy in 1962 to beat the Russians in the space race. When he came to power in January 1961, Kennedy did not plan to increase the resources of space research, but the launch of the first man in space by the Soviets[2] convinced him that it was necessary to strike hard by investing in a very ambitious space programme to save the prestige of the US. On 12 September 1962, before more than 30,000 people gathered at the Rice University Stadium in Houston, Texas, Kennedy made a famous speech:

> We choose to go to the Moon in this decade and do the other things, not because they are easy, but because they are hard, because that goal will serve to organize and measure the best of our energies and skills, because

> that challenge is one that we are willing to accept, one we are unwilling to postpone, and one which we intend to win, and the others too.

Go to the Moon? This is madness! Going to the Moon had never been done. It may be impossible! Kennedy obviously had no clear idea of all the problems that needed to be solved, but he decided that before the end of the decade, Americans would go to the Moon, and he received enthusiastic support from elected representatives from all walks of life and from public opinion, worried by Soviet successes.

A budget multiplied by 15, and nearly 400,000 people mobilized

The first budget of the new programme, called Apollo, was voted for unanimously by the American Senate. In five years, the funds allocated to NASA would be multiplied by 15! Assassinated in 1963, JFK never saw Neil Armstrong and Buzz Aldrin land on the Moon on 21 July 1969, but the programme was a real success. It benefited from a huge total budget (the equivalent of US$170 billion today) and mobilized up to 400,000 people!

School for all French people in the Third Republic, Roosevelt's Victory Program and men on the Moon – these are examples of programmes of a scale that had never been imagined, which have been completed successfully. They are also three examples of clear political decisions that: made possible what seemed totally impossible a few weeks earlier; brought millions of citizens together behind the programmes; created a dynamic that overcame all the difficulties – and there were many of course – which were resolved by thousands of brains working together; and galvanized and directed the new energies they had created.

Who can believe that people could succeed in such programmes but not be able to resolve the problem of climate deregulation? Who can believe that one country could commit tens of billions of dollars and create 400,000 jobs to go to the Moon but that we are not able to make a comparable effort for a much more fundamental purpose today: to maintain good conditions for life on Earth?

SEVEN

'€1,000 Billion for the Climate?' If It Is Really Needed, Yes, We Can Do It!

To win the war against climate deregulation, we must obviously be able to finance this colossal project. For the moment, in all our countries, we meet a 'financial cliff' that prevents the best wills in the world from deploying actions at the level required.

The problem is the same everywhere. In Germany, in November 2017, it was the industry bosses who released the results of a year's brainstorming on this issue. This is good news in itself, as it is the heart of the German economy that has got moving to succeed in the ecological transition: "It's a moral obligation and an economic opportunity", explained a spokesperson for the Bundesverbands der Deutschen Industrie (BDI).

After reading the executive summary, we understand that 'the German economy will emerge stronger from the implementation of the ecological transition' but that the country will be able to meet its commitments to reduce greenhouse gas emissions only at the cost of a 'considerable additional effort'. An 80 per cent reduction of CO_2 by 2050 should be achievable at the cost of €1.5 trillion over 30 years, or €50 billion per year; a 95 per cent drop in emissions would cost €2.3 trillion, or €75 billion a year. So, the captains of industry turned to Angela Merkel's government to see how it would be possible to finance this €50–75 billion a year.

The question of financing the ecological transition is particularly acute in the Netherlands, where the latest studies announcing an acceleration of the melting of ice in Antarctica and Greenland have been taken up by all the media. A total of 9 million Dutch people live in flood-prone areas, where 70 per cent of the country's economic activity is concentrated. If the melting of the ice continues and if the rise of the oceans accelerates,[1] much of the country risks being submerged by the waters. The climate issue dominates political life there: in early 2018, ecologists won elections in municipalities in the 12 largest cities of the kingdom. However, the question of financing the ecological transition and the financing of the adaptation and protection policy (reinforcing and raising the dikes) is as complicated to solve as in Germany.

In France, Nicolas Hulot announced a hydrogen plan to accelerate the development of this sector. We saw earlier that hydrogen is a very interesting way to improve the storage of non-controllable renewable energies (solar and wind). The plan prepared by Nicolas Hulot and his team is the result of several months of work. Alas, when it was made public, it received a chilly reception: the Ministry of Ecological Transition announced a budget of €200 million, while there is general agreement that €5–10 billion are needed. The €200 million is 25 times too little, and the Cabinet of the Minister of Finance said that this €200 million is not yet totally certain.

Similarly, all those who work on energy efficiency despair at having the means to insulate all buildings. We need to start with the 'thermal sieves' mentioned each winter because of the deplorable conditions in which millions of our fellow citizens live in Europe.[2] "To insulate a home properly, you need between €25,000–30,000", says an expert from the Abbé Pierre Foundation:

> 'To insulate 700,000 homes a year, we need €17–21 billion each year. Since we do not have such a budget, we cannot realize the objective, and to raise the numbers, we do cheap, low-quality insulations that do not do

> much good and sometimes discredit our actions... it is widely recognized that insulating houses properly would be a highly job-creating investment and very worthwhile in the medium term, but nobody can afford to invest enough.'

For France, the budget needed to finance the entire ecological transition is in the range of €45–75 billion per year (that is, 2–3 per cent of gross domestic product [GDP]), in addition to existing investments, according to I4CE, a think tank close to the Caisse des Dépôts. Half of these buildings are in the public sector and the rest are private (owned by businesses and individuals). Unfortunately, the budget just announced for 2019 is nothing like what is needed.

The reasoning is the same in all countries and in all sectors: these investments are essential and will be profitable in the medium or long term, but who will start the pump? Since we cannot find a solution at the national level, we need to seek solutions at the European level.

€1,115 billion a year, according to the European Court of Auditors

In total, for the whole of Europe, how much will the fight against climate change cost? In all honesty, we recognize that such estimates are difficult to make. The figures vary between 2 per cent and 5 per cent of European GDP, which reveals the difficulty of the calculation, as well as the vagueness of the objectives[3]: are we trying to implement only the COP21 commitments – which would lead to a warming of 3.5 °C – or do we want to finance a real revolution, like the 'zero emission' target in 2050? As long as this essential question is not settled, it is not surprising that very different figures are being thrown into the public debate. With a very ambitious scenario in mind, the European Court of Auditors estimates the necessary investments at €1,115 billion a year.

€1,100 billion a year! By listing all the public and private investments needed to make up for lost time in the fight against climate change, the European Court of Auditors has arrived at this colossal figure. For once, it may seem a little exaggerated, but as it comes from the European Court of Auditors, no one can reject this figure a priori or say it comes from fantasists who are not aware of budget constraints. Auditors at the European Court of Auditors are very familiar with these constraints, but they understand that there are higher level constraints that will shatter our businesses if we do not finance a historic climate programme.

This figure of €1,115 billion has the merit of providing an order of magnitude; it indicates that we must think outside

Box 7.1: The European Court of Auditors criticizes the ineffectiveness of the EU's climate policy and demands €1,115 billion a year

While recognizing their importance, the European Court of Auditors emphasizes the ineffectiveness of many measures in the Community's climate and energy program. According to the report published by the European Court of Auditors on Wednesday 20 September 2017, the 28 countries will have to invest €1,115 billion annually between 2021 and 2030 to achieve their objectives in the fight against climate disruption.

Without rapid decarbonization of the world economy, the global climate is expected to warm up by 3 or 4°C by the end of the century. 'In the status quo scenario, if no adaptation measures are taken, the projected climate change by 2080 would cost €190 billion per year, at constant prices, for households across the EU, which is almost 2% of the Union's current GDP', the auditors say.

The EU is committed to spending 20% of its budget on climate actions...

The problem is that this manna is still not funded. A detail.

De Tannenberg, V.L. (2017) *The Journal of the Environment*, 22 September.

the box and innovate if we want to win this fundamental battle. When Kennedy decided to go to the Moon, the US multiplied the NASA budget by 15. If we want to save planet Earth for humans, we must make a comparable effort: €1,100 billion a year. If such investments are necessary, how can such sums be found?

In 2012, when we created the 'Roosevelt Group' with Michel Rocard, Stéphane Hessel, Edgar Morin, Cynthia Fleury and a few others, we said: 'To save the banks, we put in €1,000 billion. We must put in €1,000 billion to save the climate.' Some agreed with the parallel, saying: 'Saving the banks was necessary, saving the climate is just as crucial!' However, others said that we were utopians: 'The European Central Bank will never want to create money to save the climate. What it did in 2008 was quite exceptional. It will not want to do it again and the Germans will not want it either.' So, we hit a brick wall.

In 2014, the European Central Bank turned the table upside down

However, in November 2014, Mario Draghi decided to turn things round and announced, *urbi et orbi*, that the European Central Bank (ECB) would create €1 trillion: 'The ECB will create €1 trillion to save growth. The ECB is ready to inject massive sums to help Europe's ever fragile growth. Super Mario turns the table upside down!' The news was on the front pages of all the newspapers, and some observers were astonished because the 'Germans', supposedly hostile, said nothing (see Figure 7.1).

In 2013, GDP growth in France was almost zero. In 2014, the Eurozone as a whole was close to recession. Mario Draghi wanted to do everything possible to give as much oxygen as possible to the European economy and decided to launch a vast operation of 'quantitative easing'. This was not the exceptional operation to avoid a financial collapse of 2008, but a regular and long-term monetary creation 'to save growth'.

Figure 7.1: Front page of ***Les Echos***, 7–8 November 2014

Les Echos

LE QUOTIDIEN DE L'ÉCONOMIE // VENDREDI 7 ET SAMEDI 8 NOVEMBRE 2014 // LESECHOS.FR

1.000 milliards d'euros pour sauver la croissance européenne

- La BCE est prête à injecter des sommes massives.
- L'Allemagne annonce une rallonge de son plan d'investissements publics.

Une Allema
les 25 ans d

Source: *Les Echos*, 7–8 November 2014.

We used to talk about 'printing money'. Today, there is no need for banknotes, just a few computers in the ECB offices in Frankfurt and in the headquarters of the national central banks. To make it a bit more chic and mysterious, one talks about quantitative easing, but in essence, things are very simple: the ECB creates a huge amount of money and makes it available to banks.

At the end of 2014, Mario Draghi announced that the ECB would create €1 trillion: €60 billion every month, which would be provided to private banks in exchange for the public debts they bought from the member states. The more cash that is provided to the banks, the more they would commit to reviving the European economy, the leaders of the ECB hoped. The

Box 7.2: The ECB starts buying the debts of businesses

This is a minor event for the markets: Wednesday, June 8, the ECB began to buy corporate bonds. This measure will complement the redemption of public debt that the ECB has been conducting for more than a year. In total, these purchases of securities – public and private – come to 80 billion euros each month. With one objective, boost inflation, credit and growth in the Euroarea.

Which companies will benefit? Mainly large groups. However, economists point out that they have little difficulty in finding money. The problem is rather for very small enterprises and small enterprises which, for their part, have little access to the bonds market.

Charrel, M. (2016) 'Économie', *Le Monde*, 8 June.

operation began in April 2015. A year later, in June 2016, the ECB was still creative and announced that it would not only buy bonds for banks, but also lend directly to large companies.

Air Liquide, Axa, Sanofi, Saint-Gobain, Heineken, Pernod-Ricard, Petrol d.d. and so on – the list of companies that benefited from this policy is available on the ECB's website. It is quite surprising that while thousands of small and medium-sized enterprises (SMEs) fail in Europe each year due to a lack of liquidity or equity, and while governments cannot finance the fight against climate change, we discover that Axa and LVMH, which announce billions of euros in profits every year, need the reinforcement of the ECB.

At the end of 2014, Draghi claimed that he was going to 'create €1 trillion', but in September 2016, when this symbolic threshold was crossed and while all observers noted that the effect on growth in the Eurozone remained rather weak, the ECB decided to continue this policy of quantitative easing, as well as the use of another weapon of 'unconventional' policy: loans with negative rates. Loans with negative interest rates? There is little talk of it, but it is another

Box 7.3: €233 billion borrowed in one day – the banks are onto the negative interest rate loans of the ECB

The last negative interest long-term loan from the European Central Bank attracted twice as many requests as expected. It was one of the emblematic measures of the anti-deflation mechanism of the European Central Bank: the very long-term loans with very advantageous rates granted to the banks which undertake to make credit to the economy (or 'TLTRO' i.e. Targeted Long-Term Refinancing Operations in the monetary jargon).

This device was started in September 2014. It ends on Thursday. For this last operation, the banks borrowed €233.5 billion from the ECB... An amount twice as high as expected. A total of 474 banks requested liquidity at a rate of 0% for institutions that do not respect their reimbursements, to –0.4%, i.e. the deposit rate of the ECB. In this last case, the banks make money by borrowing from the ECB.

Couet, I. (2017) *Les Echos*, 24 March.

important innovation of the ECB. When you and I have to borrow money from the bank to buy a car or have some work done on our apartments, we pay 'normal', that is, positive, interest rates, for example, of 2 per cent or 3 per cent. If we borrow €100, we have to pay back €102 or €103, and even more with interest that accumulates over several years. To help the banks, the ECB innovated, making available negative interest rate loans! 'Banks earn money by borrowing from the ECB', *Les Echos* underlines: 'They are rushing for' these TLTROs. The constraints imposed on banks are very slight: it is enough if the outstanding loan it granted increased by 2.5 per cent for a bank to benefit from negative rates. Even if it uses most of the money received from the ECB for other uses? Yes! And if the bank has not increased its outstanding loan, it only needs to pay a zero rate. We have seen more restrictive sanctions before!

Table 7.1: The LTROs issued by the ECB between September 2014 and July 2017

Ref.	Type	Settlement date	Maturity date	Days	Allotted amount
20170066	MRO	12/07/2017	19/07/2017	7	7.1 bn
20170062	LTRO	29/06/2017	28/09/2017	91	2.67 bn
20170051	LTRO	01/06/2017	31/08/2017	91	3.05 bn
20170040	LTRO	27/04/2017	27/07/2017	91	1.47 bn
20170028	LTRO	29/03/2017	24/03/2021	1,456	233.47 bn
20160133	LTRO	21/12/2016	16/12/2020	1,456	62.16 bn
20160103	LTRO	28/09/2016	30/09/2020	1,453	45.27 bn
20160063	LTRO	29/06/2016	26/09/2018	819	6.72 bn
20160065	LTRO	29/06/2016	24/06/2020	1,456	399.29 bn
20160026	LTRO	30/03/2016	26/09/2018	910	7.34 bn
20150125	LTRO	16/12/2015	26/09/2018	1,015	18.3 bn
20150097	LTRO	30/09/2015	26/09/2018	1,092	15.55 bn
20150065	LTRO	24/06/2015	26/09/2018	1,190	73.79 bn
20150034	LTRO	25/03/2015	26/09/2018	1,281	97.85 bn
20140242	LTRO	17/12/2014	26/09/2018	1,379	129.84 bn
20140189	LTRO	24/09/2014	26/09/2018	1,463	82.6 bn

Notes: MRO = main refinancing operations; LTRO = long-term refinancing operations.
Source: Created using data from the ECB.

€2,500 billion created in two and a half years

In total, according to the ECB's website, more than €1,200 billion were made available to banks via long-term refinancing operations (LTROs) or TLTROs (see Table 7.1). Let us make a quick assessment:

- with *quantitative easing*, the ECB provided private banks with some €1,450 billion; and
- with TLTROs, the ECB provided private banks with just over €1,200 billion.

All in all, and without any other operations that we have not been able to identify,[4] a little over €2,600 billion has been made available to banks in less than three years by the ECB. Including prepayments made by some banks, the net balance of quantitative easing in the broad sense (that is, quantitative easing plus TLTROs) amounted to €2,500 billion at the end of 2017. It is colossal and is hardly ever mentioned in the public debate: between April 2015 and December 2017, the ECB created, *ex nihilo*, and made available to banks and some large companies some €2,500 billion – more than all the wealth created in France every year!

Let us go back to the fundamental purpose of this little book: to get the implementation of an operation to fight against climate change on the scale of the Marshall Plan. In the spring of 2014, some people told us '€1,000 billion for the climate? It is not possible. The Germans will not want it. The ECB will not either.' However, four years later, we see that the ECB has made available to banks more than €2,500 billion in two and a half years, or *€1,000 billion per year*, and the German leaders allowed this!

Some 2,500,000,000,000 euros – a huge sum! In all our countries, we cannot find substantial financing to fight against climate change, but given these figures, it is difficult to say that the funds do not exist.

Where did these colossal sums go? As the ECB has announced '*targeted long-term refinancing operations*', it is fair to check that the targeted 'targets' have been achieved and that these loans have been directed to long-term investments that are useful for the common good or, *at least*, for economic development.

New money for investment: 11 per cent of the sums created by the ECB

One can see all the credit distributed by private banks over the period from mid-2015 to the end of 2017 on the ECB website. While the ECB made €2,500 billion available to them, the new loans distributed to industry by these banks over this period represented €280 billion, or 11 per cent of the colossal sums made available by the ECB. Where did the rest go? Essentially, it has gone to feed speculation on the financial markets. Since 2014, there has not been one month when the major global financial markets did not break new records: in 2008, the Dow Jones peaked at 14,000 points before collapsing to less than 7,000 points in 2009. Since the central banks started *quantitative easing*, every day has been a party on Wall Street: the Dow Jones reached 17,000 points in the spring of 2014, then 19,000 points in 2016. The threshold of 20,000 points was crossed in January 2017, and in early 2018, the Dow reached the 25,000 mark – more than a 230 per cent increase from the 2009 low point (see Figure 7.2)!

Donald Trump exulted: '*Great!* Business is better than ever.' He is a worthy heir to President Hoover, who declared in September 1929, a few days before the biggest crash in history: 'Prosperity is around the corner.'

Trump exults, but the International Monetary Fund (IMF) keeps sounding the alarm. In Davos, in February 2018,

Figure 7.2: The Dow Jones index from January 1995 to January 2018

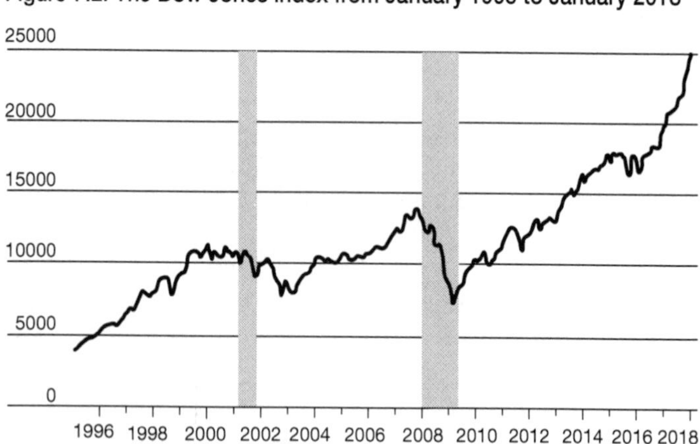

Christine Lagarde warned that there was a risk of a crisis 'more serious, faster and more general than in 2008'.

The whole planet is at risk of a new crisis, worse than the one of 2008. A real economic crisis, not just a financial crisis, because, in reality, there are two distinct problems, which feed each other:

- a level of speculation never seen before; and
- a level of debt never seen before.

If the financial markets were the only problem, it would not be too bad. If markets fall back into their past mistakes but the fundamentals of the economy are solid, we can rest easy ('they do what they want, but they will pay the consequences of their lack of professionalism'). Alas, this is not the situation at all: the entire global economic system is increasingly unbalanced – and this time, we cannot say that we did not know because the IMF, the World Bank, the Bank for International Settlements (BIS) or the Institute of International Finance (IIF) have been giving warnings for seven years.

Unprecedented alert on debts

'An unprecedented warning about the debt of Western economies' headlined *Le Monde* already in the spring of 2011: 'The United States has regained some growth, which benefits all Western countries through their exports, but at what price? In only ten years, the public debt of the United States has nearly doubled.' 'The systems often last longer than we think, but they eventually collapse much faster than we imagine', worries Kenneth Rogoff, the former chief economist of the IMF, who sees the levels of debt on all continents rising without any real attempt to calm speculators or to attack the roots of the crisis: why do all our countries need more and more debt to avoid recession?

Figure 7.3: Front page of *Les Echos*, 13 April 2016

Les Echos

LE QUOTIDIEN DE L'ÉCONOMIE // MERCREDI 13 AVRIL 2016 // LESECHOS.FR

Le FMI redoute une nouvelle crise mondiale

Source: *Les Echos*, 13 April 2016.

'The IMF fears a new world crisis', *Les Echos* warned again in April 2016, explaining that the US is over-indebted, but the situation is not brighter in Japan or in China either (see Figure 7.3). Do our leaders not read the papers? Not a month goes by without an official body issuing an alarming report on the economic and financial situation of the planet, no doubt trying to force central banks and heads of state to change their policies – in vain so far.

In July 2018, the IIF published a damning balance sheet of the global financial situation. The IIF brings together almost all major international banks.[5] Gathering information from all continents, the IIF reaches frankly disturbing conclusions. A summary of the key figures appeared in *Le Figaro*.

Global debt (private and public) has increased by US$150 billion in 15 years and has reached 318 per cent of world GDP

- Total global debt stands at US$249 trillion at the end of the first quarter of 2018, or 318 per cent of the world's GDP.
- It has increased by US$29,000 billion since the end of 2016 and US$150,000 billion in 15 years.

Box 7.4: China's debt risks exploding says the Bank for International Settlements

China's total debt amounted to 168,480 billion yuan ($25 trillion) by the end of 2015, i.e. 249% of her GDP.

The 'difference between China's credit-to-GDP ratio and its long-term trend' reached 30% in the first quarter of 2016, a worrying level never seen before, shows a quarterly report released Sunday night by the Bank for International Settlements.

China's debt may explode over the next three years, warns the Central Bank of central banks.

Le Figaro (2016) 19 September (lefigaro.fr with AFP).

- Financial sector debt has reached a record high of US$60.6 trillion.
- Non-financial corporate debt is at record levels in Canada, France and Switzerland.
- In the US, the government's debt exceeds 101 per cent of GDP.
- In emerging countries, debt reached a record level of US$58,500 billion.

Since the 2008 financial crisis, debt has not declined worldwide; rather, it has increased by 40 per cent (see Table 7.2). Moreover, with the current rise in interest rates, the levels of debt of all structures risks 'destabilizing the world economy', as the World Bank says modestly.

The IIF provides us with a splendid chart that identifies the most threatening bubbles in the global economy (see Figure 7.4). The two countries often presented as 'the two engines of the global economy', the US and China, are actually two huge debt bubbles, two time bombs.

Table 7.2: Changes in world debt by major sectors between 2008 and 2018

	2008	2018
Households	37	47
Businesses – non-financial	46	74
Debts – public	37	67
Debt – financial sector	58	61
Total	178,000 billion	249,000 billion

Notes: In billions of dollars. For both years, the total debt is compared at the end of the first quarter. These are the latest figures available for 2018 at the time of writing this book.
Source: IIF, July 2018.

Figure 7.4: Private sector debt (non-financial) – the US, China and Japan are the main sources of financial risk

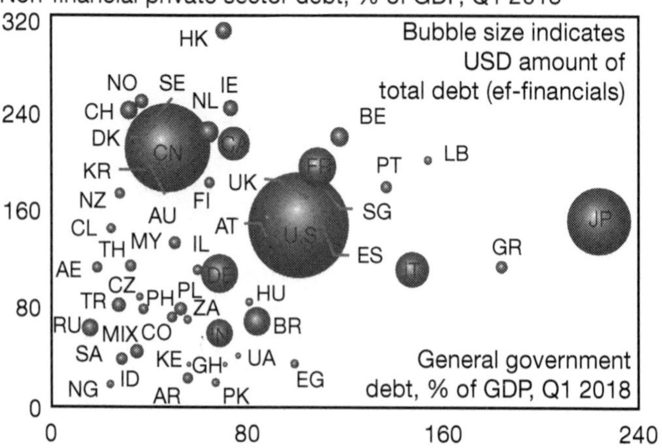

Source: IIF.

Table 7.3: A balance sheet for the US, 2017

Increase in GDP	+427 billion US dollars
Increase in debt	+1,811 billion US dollars

Source: Data from the Federal Reserve, August 2018.

In the US, household debt has exceeded the peak of 2008.[6] Corporate debt is also at a level that has never been seen before. Public debt has increased fastest; with the tax giveaways offered by Donald Trump to shareholders and the wealthiest households, the public deficit has exploded. The government's debt has just exceeded US$20,100 billion, and the forecasts for the next few years are chilling.

In total, at the end of March 2018, US debt reached US$49,831 billion.[7] For those who marvel at the growth rate of the US (+2.3 per cent in 2017), it would be wise to compare GDP growth with debt growth (see Table 7.3).

In 2017, it took 4.2 extra doses of debt to get one dose of growth. The figures for the beginning of 2018 are even more disturbing: the tax reform passed by Donald Trump and his allies started to really increase the public deficit, with total debt increasing by US$874 billion in just three months, including US$630 billion of additional public debt.

For some months now, some commentators have been saying, 'Certainly, one can criticize Donald Trump's methods, but in the US, at least it has results.' They have missed the point and have forgotten everything about the 2008 crisis! Admittedly, the growth of the US is a little stronger than that of Europe, but with such growth of debt (+US$874 billion of debt in only three months in early 2018!), one cannot say that 'Trump has had results'; rather, Trump is taking the US into a dead end – and the whole planet with them if we do not quickly find a new model for prosperity to protect us as much as possible from the coming crisis and to launch an autonomous dynamic of job creation. Like a car that needs a litre of engine oil every 300 yards and can blow up its engine

Figure 7.5: China, total debt (total credit plus total outstanding amount of bonds, as a percentage of GDP)

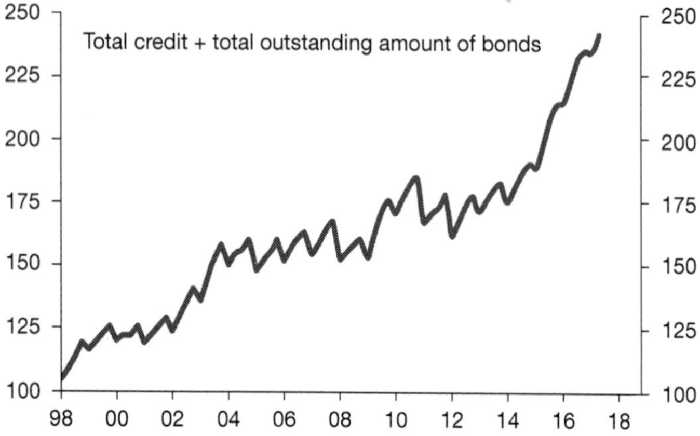

Sources: Datastream, BRI, NES, NATIXIS.

at any time, the US economy continues to grow, of course, but with a ratio of debt to growth that is 'totally explosive'. As for China, its total debt now exceeds 250 per cent of GDP, and doubtful credits have never played a bigger role in its economy. The Chinese government keeps saying that it wants to calm the debt surge, but the figures published in August 2018 show that it is not getting there: in the first half of 2018, total new loans grew at an annual rate of +13 per cent![8]

Beyond the amount, the most worrying thing is probably the shape of the curve: the debt of the US and China is increasing even faster than before 2008 (see Figure 7.5). Joseph Stiglitz, a Nobel Prize-winner in economics, has said that we have not solved any of the problems that led to the 2007–08 crisis: 'We just moved the chairs on the deck of the *Titanic*.'

A brutal financial crisis

What are our governments doing? How is it that these questions are never debated in our parliaments? In 2008, a crisis born in the US crossed the oceans in a few days and caused tens of millions of people on every continent to become unemployed. What will happen when the bubbles break in China and the US? More and more voices are criticizing the role of central banks since 2017. Can '*pyromaniacs*', as Patrick Artus[9] calls them in *Les Incendiaries*, adapt their policies to changes in the global economy, or are they prisoners of dogmas from the 1960s or 1970s? 'More and more, central bank statements and actions are out of step with the expectations of citizens and the mechanisms that govern contemporary economies', writes Patrick Artus. Central banks were created to facilitate the smooth running of economies, ensuring that there is just enough money in circulation and avoiding financial crises by monitoring risks. Today, they are failing all down the line: they behave like pyromaniac firefighters.

They continue to think as if they have to manage closed economies, where any increase in credit leads to increased demand and employment. However, with globalization, increasing use of robots and the financialization of the economy, this is no longer the case. Patrick Artus rightly calls for less independent central banks, which should decide on the objectives of monetary policy with governments and parliaments.

'The worst is before us'

'My experience as a banker and regulator of the financial markets leads me to say that the worst is unfortunately ahead of us', explains Jean-Michel Naulot, former member of the Council of the French Financial Markets Authority:

> We had a glimpse of what could be a collapse with the 2008 crisis, but the risks of a real collapse are still here. The worst is before us. If our leaders refuse to examine their approach, if they want to *stay on this tack* and continue to make speeches as though reciting a catechism, they will take us straight into a cul-de-sac.

As the person who was one of Michel Barnier's close colleagues when he was European Commissioner goes on to explain:

> By their obstinacy, it is they who are driving populism! And they have the audacity to tell us that if there is a crisis, it's because they have been prevented from doing more! The financial crisis is inseparable from a much deeper crisis, that of astonishing economic liberalism which has run out of steam. It is not by correcting the financial system marginally that the crises will be stopped. The responsibility of leaders is obvious: to make moderate speeches, as most politicians do, when we are facing a situation of exceptional gravity, is to risk a future collapse.[10]
>
> For these leaders, financial crises are inherent in the capitalist system. Crises are 'the price to pay'. Are they really interested in this subject? Have they understood the seriousness of what is on the way?[11]

Financial markets where speculators defy the laws of gravity, US debt at a record level and Chinese debt out of control – 'The next crisis may be more serious than that of 1930', said Mervyn King, Governor of the Bank of England in 2012. However, none of the imbalances he denounced has been addressed since – quite the contrary!

The comparison with 1930 is fair: did the crisis of 1930 not also come after a previous crisis in 1920–21? This first financial crisis was quite severe but short-lived, and in a few years, it was forgotten: confidence returned, which gave way to euphoria, and little by little, many investors believed that they could get away with anything and forget the basic rules of prudence, that is, until the great crash of October 1929. Ten years after the first crash, the crisis of 1929 led to a surge of unemployment and upheaval. History is never exactly repeated, and the responses of central banks at the beginning of the 21st century are not the same as in the first quarter of the 20th, but no one can deny today that we are playing with fire.

One does not need to be a Nobel Prize-winner to understand that this development 'model' is by no means sustainable: it is not sustainable ecologically (as we saw earlier) and is not sustainable economically either. Ecological debt and financial debt are two facets of the same flight forward, an increasingly unbalanced system that is leading us to the abyss.

'A new crisis could cause, calculates the IMF, ten times more damage than ten years ago', *Les Echos* insisted again in April

---

Box 7.5: The world economy is like the Titanic: it speeds up before the shock

The warning comes from the IMF: the world economy looks like the Titanic before it sank. Like the Titanic before the collision with an iceberg, the global economy is accelerating: the International Monetary Fund has revised its growth forecasts upward, suggesting a sustainable recovery in many countries.

But the ship's hold is more and more weighed down by debt. 'The waters seem calm, but problems are accumulating below the surface and threaten to torpedo the global recovery and lead to a global financial crisis like in 2008.'[13] How can one explain this? There is too much money in the financial system. Central banks have injected masses of funds into the economy. This has ended up by creating the risk of financial bubbles.

There is now too much debt in the world: the G20 countries are dragging a ball and chain of debt that represents more than two years of their GDPs. If interest rates rise, and they risk doing so, these countries will be strangled. And the attempt by China to flee from the problem by creating even more debt reassures no-one. The result is an explosive cocktail that justifies the IMF's warning: 'If nothing changes, we will relive the crisis of 2008.' We have been warned.

Barré, N. (2017) 'Europe 1' (economic editorial of the Director), *Les Echos*, 12 October.

2018.[12] We cannot say we did not know. In this area too, 'it will soon be too late'.

Homo sapiens sapiens or *debilus debilus*?

Before presenting our solutions to get out of the crisis, let us run quickly through the main points of what we have just described. Each of us probably feels deeply disturbed by what is happening: we were taught in school that we are *Homo sapiens sapiens*, the most advanced of the primates. In view of the crises that we provoke and the way we suffer without reacting, we may wonder if we are really *sapiens sapiens* and not rather *debilus debilus*.

Like the rabbit frozen in the headlights of the truck about to crush it, we do nothing when, every month, extreme weather events make headlines, climatologists tell us that we have only a few years to react and the IMF, the World Bank or the IIF alert us to the seriousness of the financial crisis that threatens. The money that would finance the ecological transition exists, but instead of settling these major, linked issues, they get worse and worse, fuelling speculation and boosting inequalities before the final collapse. If nothing changes, the story is easy to imagine: we will soon suffer a new financial crisis, with soaring unemployment and populism, and an irreversible aggravation of climate change – and we will not be able to say that we did not know!

Is this how people are supposed to live? Are we stupid enough to accept being tossed from one crisis to another, unable to take control of our future? When we say that it is the future of humanity that is at stake (and not that of the planet), this is true for two reasons: first, because of the hundreds of millions of people whose lives will be upset – or shattered – by climate change; and, second, because in the depth of each of us is this desire to protect humanity, an anger and fierce desire to refuse the nonsense of a world that is about to collapse.

'When I respect the sacredness of the world around me, I respect my own sacredness', said Paul Ricoeur: by destroying the planet that hosts us, we are actually destroying something sacred in us. 'Still more money for banks, but no money for the common good, and climate! If we wanted to revive terrorist movements, we could not do better', lamented the leader of a very large German bank a few years ago. He was right.

Set fire to one's own house – how stupid! It is time to reject these absurd principles that govern our time and lead us to the abyss, refuse to accept climate change, famines and wars, and refuse to accept the nonsense in the world around us – a world that has never been so rich, where liquidities flow, where shareholders do not know what to do with their dividends and where unemployment, precariousness and poverty cause so much suffering.

EIGHT

Putting Finance Back at the Service of the Common Good: The European Climate Finance Pact

To stop the chaos that is on the horizon, we need to reduce speculation and radically change gear in the fight against climate change. This is why we ask that the EU acquires a very powerful tool to finance the ecological transition in Europe, in Africa and all around the Mediterranean.

Just as there are hybrid cars, with two power sources, we propose a hybrid climate pact, with two sources of funding. To put monetary creation at the service of the fight against climate change, we first want to create a European Climate Bank: a subsidiary of the European Investment Bank (EIB). From its birth, the European Climate Bank would benefit from the AAA rating of its parent bank, and it would be charged by its statutes with solely financing the ecological transition. Each country would have the equivalent of 2 per cent of its GDP as loans at 0 per cent interest rates for 30 years from this European Climate Bank; therefore, France would have €45 billion a year available for private and public investments, Germany would have €60 billion each year, Poland €16 billion and so on.

Two per cent of GDP each year for 30 years

As mentioned earlier, this figure of '2 per cent of GDP' is a reasonable estimate of the funding required for the ecological transition, in addition to what is already available. This is the estimate made by Lord Nicholas Stern; it is also the amount proposed by the experts at France's Caisse des Dépôts. In reality, this figure is only indicative: in all countries, it is necessary to launch programmes (in building, transport, renewable energies, agriculture and so on) and to tackle the problem of available skills. 'Obviously, if tomorrow, €30 billion appear from nowhere to finance the insulation of buildings, we would be unable to use it because we would lack skilled craftsmen', says an official of the French Building Federation:

> So, in the first year or two, we would not be able to utilise that 2% of GDP. But if we know that we will have this budget for thirty years, and if, as proposed, the insulation of buildings is made obligatory in twenty or thirty years, that changes everything. This is a real revolution, and in each country, in each region, we would train people in building, in renewable energy and in transport. And, by taking a few years to pick up speed, *making sure of the quality of what is done*, we could invest much more than 2% of GDP.

This allocation of 2 per cent of GDP would be used to finance investments, both private (the insulation of our homes, factories and offices) and public (public transport and the insulation of schools and town halls).

As Alain Grandjean and Gaël Giraud regularly point out, it is important to take these public investments out of the calculation of the public deficit (the 3 per cent of the Maastricht Treaty):

> It is urgent to set up a major investment program in favour of the energy and ecological transition. In the current context of strong budget constraints, it is necessary to take these investments out of the calculation of the public deficit, as would any private enterprise that does not confuse its investments with its operating expenses.[1]

How, then, do we decide which public investments remain in the deficit calculation and which ones are excluded? A simple rule could be that public investment in the fight against climate change financed by the EIB is, a priori, excluded from the calculation. In each country, the parliament would decide on the investments financed by these loans at the 0 per cent rate.[2] It would be up to each country to publish a summary of these investments each year, allowing a high degree of transparency, an exchange of good practices and an a posteriori audit by the European Court of Auditors.

Create a European bank dedicated to climate

At the end of 2017, when we relaunched the debate on a European Climate Pact, we proposed creating a European Climate Bank by transforming the EIB. Among our supporters was Philippe Maystadt, who was the President of the EIB for 11 years and its Honorary President until December 2017.[3] No one could doubt the seriousness of our proposal. However, in the spring of 2018, during a debate organized in Brussels with officials of the EIB, ECB and Bank of France, leaders of the EIB, who have since decided to support our project,[4] convinced us that for the project to succeed as soon as possible, it would be better to create a new bank rather than transform the EIB. As our interlocutors explained:

> 'If you want to transform the EIB so that it is 100 per cent geared towards the ecological transition, you will run up against the lobby of companies and local councils that still need credit for investments that are not 100 per cent green. Do not worry: they will not get support for brown investments (oil, gas …), but all their investments are not green either. And in many cases, SMEs and local councils need the EIB to finance them. An EIB transformed into a 100 per cent climate bank could no longer do it. It would be a shame and they will do everything to block you.

> The EIB can be prevented from making certain climate-damaging investments, but it cannot be expected to make only green investments in the short or medium term. If your goal is for the treaty to be adopted in 2019 and to be applied in 2020, then it would be better to create an EIB subsidiary which would have the status of a bank (which the EIB does not have) and which would be 100 per cent for the ecological transition.'

In 1989, six months were enough for Kohl and Mitterrand

As an EIB official added:

> 'In 1989, after the fall of the Berlin Wall, it took six months for François Mitterrand and Helmut Kohl to create a bank to finance the transition of the former Soviet countries. Starting from scratch, it took only six months to create the EBRD and it took just one year for it to start work! For the Climate Bank, we would not start from scratch: we can rely on all the know-how of the EIB and its allies. It can be very quick. I am ready to come with you to the Elysée or the Chancery in Berlin to explain this to the people you are dealing with.'

Guaranteed by a treaty (so with the stability that a treaty gives), massive funding from the European Climate Bank would start the process and finance at 0 per cent interest rates all projects that will have positive effects in the medium or long term but that are blocked financially today. This would already be a decisive element in accelerating the ecological transition.

However, loans at 0 per cent interest rates are not enough on their own: it is time to create a real climate budget at the European level. In all the countries, there is an education budget, a health budget, a research budget and a defence budget. If we want to declare war on climate change, a proper climate budget at the European level is needed.

Finally, a European climate budget

When France decided to make schooling available to all in the 19th century, it did not offer parents loans at 0 per cent to finance their children's education; rather, 'schools became free of charge the week before being mandatory', recalls an education historian. It was said earlier that in order to win the climate battle, the question of energy efficiency is fundamental, and it was pointed out that in order to properly insulate a house or apartment, it takes on average between €25,000 and €30,000. This sum is too much to ask all citizens to finance themselves: even if they are convinced of the importance of fighting against climate change, and even if they know they would save money on their heating (and air-conditioning) costs, a large number of households would never be able to afford such a sum – and the reasoning is the same for many SMEs and communities, which would find it difficult to fund such expenditure alone.

To pay a large part of all the costs of improving energy efficiency, and to invest massively in new modes of transport, we propose the creation of a European budget. When Emmanuel Macron repeatedly asserted during his campaign and since his election that he wants 'a new ambition for Europe' and that he wants 'a European budget of several hundred billion', we say: 'Great!' For the climate, we need a budget of €100 billion a year: €100 billion that we can invest in fundamentally useful projects but that are not profitable from an accountant's point of view, like education and health improvement are absolutely useful but not, strictly, 'profitable'.

A budget of €100 billion a year? This would provide €40 billion for projects in Africa and around the Mediterranean, €10 billion a year for research, and €50 billion to pay a portion of the bills of individuals, businesses, associations and communities in all our countries. How could this €100 billion be found? It seems to us that we cannot tax all citizens more: we know that Europe is already not very popular with a growing number of its citizens, and the introduction of a new tax affecting all families would be not be well perceived. So, how can Europe raise new financial resources without burdening household budgets? The best solution is to take up an idea proposed by Jacques Delors 20 years ago: stopping fiscal dumping in Europe.

Stopping European fiscal dumping

There is very little talk of it, but in less than 30 years, competition between European countries has led to a sharp drop in the tax on company profits (see Figure 8.1). Each European country lowers its corporation tax rate on profits for fear that companies might move to a neighbouring country that has just lowered its tax rate.

At the European level, the average corporate tax rate, which was about 45 per cent in 1985, has fallen to less than 23 per cent today, and the average effective rate is less than 20 per cent, according to the European Commission. The French newspaper *Ouest France* has suggested that this sharp drop in

Figure 8.1: Average corporate tax rate, 1993 to 2010

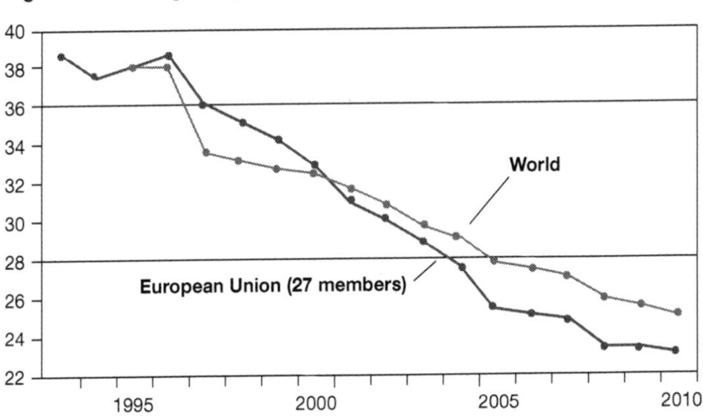

Source: KPMG.

the average rate in Europe is a consequence of the stability of the tax rate in the US: from Roosevelt to Trump, the rate was stable at 38 per cent.[5]

When Roosevelt arrived in the White House, the US suffered the same fiscal dumping: Florida lowered its tax on profits, then it was Texas or Arkansas that lowered their rates. Companies were engaging in tax tourism, and when the crisis of 1929 erupted, governments (both federal and state) found themselves without financial means to fight the consequences of the crisis. However, Roosevelt, elected in late 1932, decided to break with this logic. In less than six months, he had created a federal tax: whether in Florida or Texas, a company would pay the same contribution. Washington levies the taxes and keeps some of the funds to support the federal budget before sending the rest back to the individual states. For a few weeks, shareholders cried (wolf) and accused Roosevelt of being a communist who was going to kill the US economy. However, Roosevelt held out and the US has lived for more than 80 years with a rate of 38 per cent, while the average rate in Europe has fallen to 19 per cent (see Figure 8.2).

The world has turned upside down! A liberal country like the US has a tax rate on profits twice as high as Europe. All the alarmist speeches made in 1933 by the shareholders' lobby were wrong: the creation of this federal tax did not reduce the prosperity of the US – quite the contrary!

To finance a climate budget, what stops us from breaking intra-European fiscal competition and creating a climate contribution, for example, of 5 per cent of profits on average? 'You're right really', some say, 'but with Donald Trump, the tax rate on profits has dropped significantly in the US. This reduces our flexibility.' Donald Trump has passed a fairly radical tax reform, but the tax rate on profits remains at 24 per cent. It is therefore still possible to create a European tax of around 5 per cent without the overall rate being higher in Europe than in the US.

Figure 8.2: Changes in the corporate tax rate, the US and EU

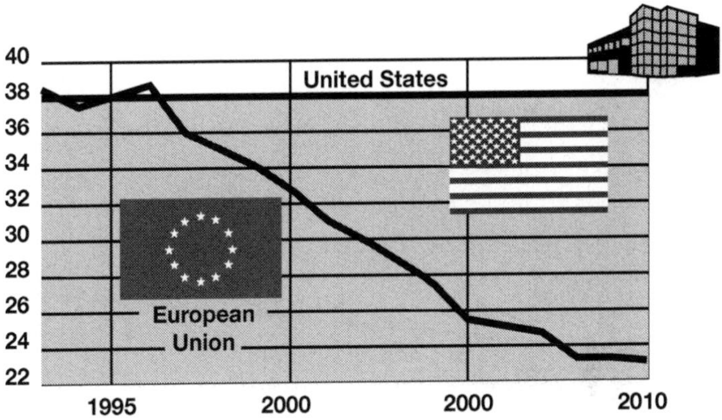

Source: KPMG.

A climate contribution? A 5 per cent tax on profits that are not reinvested

The net profits of companies in the Euro area in 2016 were around €410 billion for financial companies and €1,500 billion for other companies. A 5 per cent climate contribution would therefore bring in €100 billion each year. From a macroeconomic point of view, this contribution of 5 per cent is also justified by the fall in the share of wages in Europe's GDP (a fall of 12 per cent in 30 years!) (see Figure 8.3). Moreover, the 'corollary' of this relative fall in wages is soaring profits. The work of the IMF on the causes of the debt crisis shows that on all continents, the share of wages relative to GDP is at its lowest level in history, while the levels of dividends, which often leave the real economy to go to the financial markets (when they do not go directly to tax havens), have never been so high (see Figure 8.4). That is why more debt (public or private) is needed all the time to maintain some growth.

Figure 8.3: Proportion of wages as a percentage of Europe's GDP

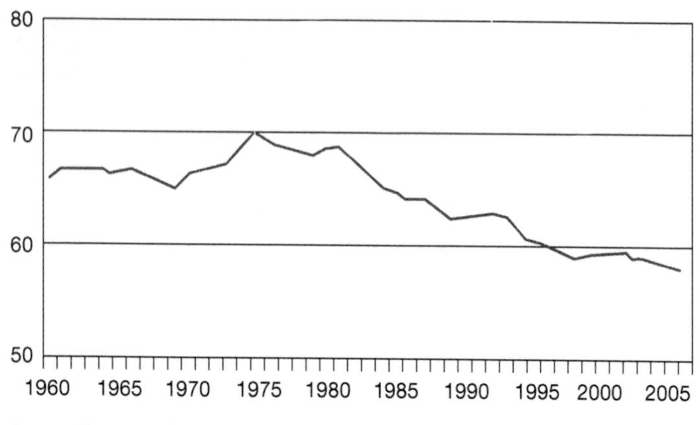

Source: European Commission.

Figure 8.4: Dividends compared to salaries (as a percentage) in French industry, 1996 to 2014

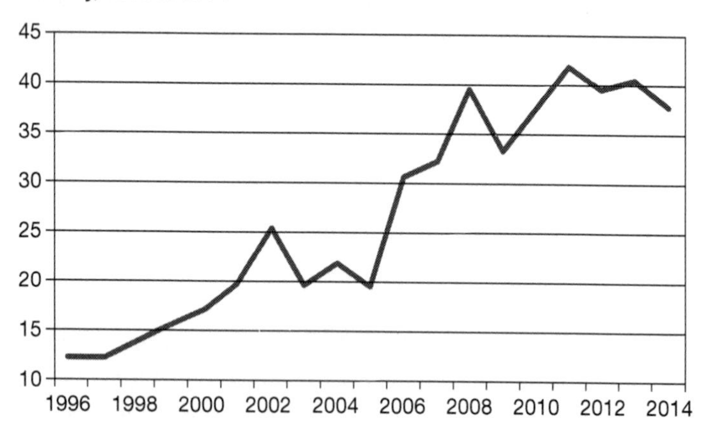

In Europe, as in all continents, a growing share of profits goes neither to investment in the companies themselves, nor to profit-sharing with employees, but to feed the financial bubble. Taking 5 per cent of the profits to beef up our investments and give better incomes to millions of Europeans would be a very good thing. This would have the effect of deflating speculation a little and strengthening the resilience of our societies: when the next financial crisis hits, its impact will be much less if we have launched, *before this crisis*, a very strong dynamic of job creation in all our countries.

"Even if I am taken for a neo-Trotskyist, I am in favour of this tax of 5 per cent", the President of a large French employers' federation said recently, with a smile:

> 'The climate question is becoming very serious. It is indeed necessary to release resources for a real budget and to take 5 per cent from shareholders is a reasonable contribution. We will require efforts from all citizens (using our cars and planes less often, eating less meat, insulating our homes, producing less rubbish, etc). It is therefore hard to see why shareholders should not also contribute to the collective effort. Especially since the fruit of this contribution of 5 per cent would also serve the companies: like all the players, they would receive help to carry out improvements, which would then allow them to decrease their energy bills in the long term. Some will scream before they are hurt, but a 5 per cent contribution is manageable.'

Let us also remember that if we create 800,000–900,000 jobs in France alone, the whole of the economy would be strengthened. In the autumn of 2018, the National Institute of Statistics and Economic Studies (INSEE) announced a slow-down in activity due to the low level of household spending, and the IMF regularly warns us about the risks of a major

Box 8.1: New record for dividends: nearly US$500 billion in the second quarter of 2018

It's a record. Listed companies distributed $497.4 billion in dividends in the second quarter, an increase of 12.9% in a year.

Businesses are doing well. Very well, according to a study by asset managers Janus Henderson. Over the course of 2018, companies are expected to pay more than $1,350 billion in dividends worldwide.

At the top of the ranking, Europe, excluding the United Kingdom, contributed significantly to this record, with $176.5 billion in dividends distributed in the second quarter. With 18.7% growth (7.5% in fact once adjusted for extraordinary dividends and currency effects), Europe has achieved its best distribution since the second quarter of 2015.

Over the period, 50.9 billion dollars were paid out in France, an increase of 23.5% in one year. Among the 12 highest-paying countries, the United States, Japan and France have the top places.

Le Figaro (2018) 20 August.

crisis; if the climate pact creates many jobs in all our countries, thousands of SMEs would see their order books fill up.

In France, –8 per cent + 5 per cent = –3 per cent

In France, the creation of a federal tax of 5 per cent on profits would be painless because the government has planned to lower the tax rate on profits by 8 per cent and, by 2022, 'to converge towards the European average'. If a European treaty creates a 5 per cent climate contribution, French companies would benefit from a 3 per cent decrease in tax instead of the 8 per cent announced – nothing dramatic.

All the major European media have featured the seriousness of climate change this year (2018). All major media have also spoken of the explosion of dividends. When dividends rise by 18 per cent in one year in Europe, it would be unacceptable for shareholders to refuse a contribution of 5 per cent on their profits. If it is because they do not give a toss about climate change, they have to say so! Thus, financed by the European Climate Bank, on the one hand, and by a climate budget of €100 billion per year, on the other, this European Climate Pact would boost actions in all our countries to achieve zero net emissions by 2050.

Split the bill in half

Thanks to this European Climate Pact, in many regions, local councils could get support for paying half the cost of the necessary investments (see Figure 8.5). In the case of upgrading insulation, for example, when each of us (households, businesses or communities) paid a bill for insulation work, we would receive a cheque for half of the sum from the member states.[6] The other half could be financed by a 0 per cent loan that would

Figure 8.5: Costs halved for individuals, businesses and communities

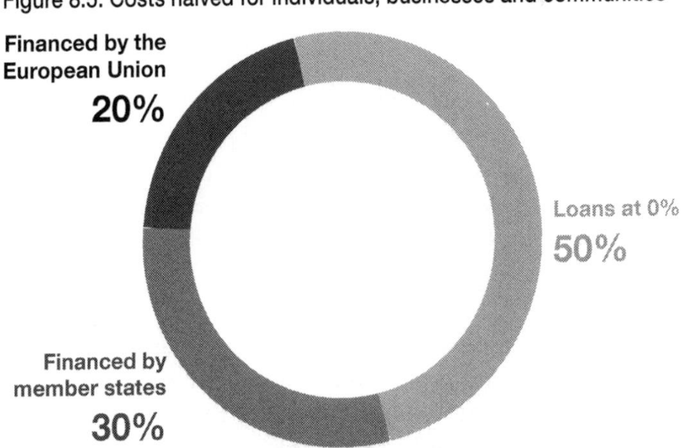

be obtained from our banks, being derived from the national envelope provided by the European Climate Bank. Then, thanks to the savings that we would realize on our heating costs, we would be able to repay this loan in 10 or 15 years.

Education for all is free. The insulation of our homes would not be totally free, but to repay the 0 per cent loans, our heating costs would be greatly reduced. This would force us to pay close attention to our behaviour and not 'leave windows wide open during the winter'. The occupants of many houses and buildings, both offices and residential, testify to the fact that we feel much more comfortable in well-insulated buildings and that they make substantial savings.

Make improvements in energy efficiency mandatory

By halving the cost and by giving ourselves the means to train competent people in all our countries to carry out the work, it would be possible to upgrade the insulation of all buildings in the medium term and to make it a rule that all public and private buildings are insulated within 20 years. "It will be necessary to start with buildings with the worst heat loss, including financial support exceeding 50 per cent for the poorest households", says an expert:

> 'And start without delay the programme for all buildings: today, the law says that no one can sell a house or a building contaminated with termites or asbestos. Similarly, one can imagine providing a thermal assessment and making insulation work mandatory when a property is sold, as it would be empty for a time.'

If the European Climate Pact could simultaneously finance the insulation of all buildings, massive investment in public transport, a change in farming methods and the development of renewable energy, the effect on employment would undoubtedly meet the Agence de l'environnement et de la maîtrise de l'énergie's (Agency for Ecological Transition's [ADEME's]) calculation: between 700,000 and 900,000 jobs created in France, and between 5 and 6 million all over Europe – and probably even more south of the Mediterranean, where it would also be necessary to invest heavily in adaptations for climate change.

There are, of course, sectors where jobs would disappear: we must ensure that all those working in the coal sector, for example, in Poland or Germany, can find a properly paid job close to home. Everyone should have access to the necessary training and be supported if needed in this retraining, which can be complicated for some. Yes, there would be significantly fewer coal miners and significantly fewer truck drivers, but we would create hundreds of thousands of jobs in the building trades, renewable energy, public transport, research, agriculture and so on. There is no doubt that the balance sheet would be very positive and that useful, profitable and non-relocatable jobs would be created everywhere.

"If we obtain this European Pact, it would change everything", so said Nicolas Hulot's advisers on farming issues in December 2017. Yes, it would change everything if Europe could give us funding at this level, guaranteed for 30 years. It would change everything in Europe and Africa, and that would undoubtedly inspire other large regions of the world to follow the same path.

However, even if it could 'change everything', this European Climate Pact is obviously not a panacea. The European Climate Pact has many advantages, and its immediate adoption would undoubtedly have a decisive effect, but it would be ridiculous to claim that the European Climate Pact alone would be enough to win the climate battle. To make up for the lost time and to avoid the chaos threatened by climate deregulation, it will be necessary to use all the levers available. "It's like fighting the AIDS virus", says a doctor friend. No single molecule can kill the virus alone. It is thanks to tri-therapy that we have made great advances: by taking three molecules at the same time, we can knock out HIV. In the same way, against the disturbance

Figure 8.6: To win the battle, we need to use all available weapons

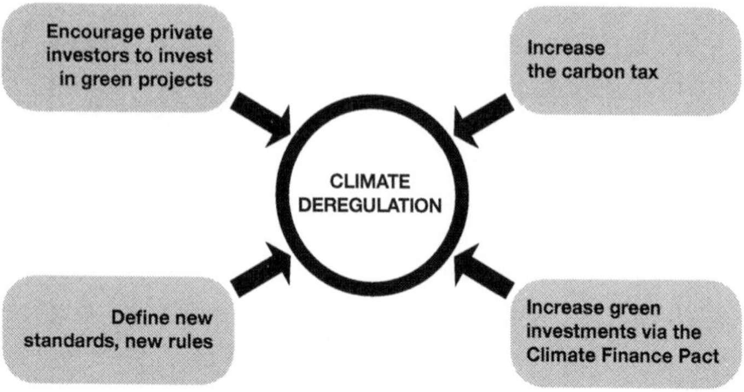

Encourage private investors to invest in green projects

Increase the carbon tax

CLIMATE DEREGULATION

Define new standards, new rules

Increase green investments via the Climate Finance Pact

of our climate, it is necessary to act with force on at least four levers (see Figure 8.6).

It will be necessary to pursue the following at the same time:

- Private financing should be directed towards green investments: stop all public subsidies for fossil fuels and very strongly encourage banks and insurance companies to stop their brown investments (oil, gas and so on) and to accelerate the redeployment of their funds towards renewable energy. The European Commission has provided a very interesting report on this. We must now go further and give legal force to its proposals.[7]
- The price of carbon should be increased so that those who do not want to change their practices know that they will pay more and more for the damage they cause. This point should be included in the European Treaty so that no one doubts the inevitability of this future increase in the carbon tax.[8]
- New rules of the game should be defined and put into law after a wide public debate: 'In 20 years, all public and private buildings must be insulated. In 20 years, no car will be allowed on the roads if it produces more than x g CO_2

per 100 km. No truck will be able to drive if it produces more than y g CO_2. In 20 years, all the farms will have to have a carbon balance and a nitrogen balance lower than …'. These new rules of the game will need to result from collective decision-making, involving all the stakeholders, and will also have important consequences for directing our research in Europe.

- Green investment should be increased through the European Climate Pact, which will make it possible to finance all these works adequately.

After having participated in a large number of public debates on this subject, it is clear to us that the Climate Pact is not a panacea, but that it can resolve a number of challenges (see Figure 8.7):

- The Climate Pact would help to keep alive a reasonable hope of winning the climate battle. If Europe and Africa

Figure 8.7: The Climate Pact could resolve a number of challenges

EUROPEAN CLIMATE PACT

Win the battle for the climate

Create jobs, massively

Reduce speculation

Provide a vision for Europe

Reinforce our policy for research

Increase people's buying power

Provide a long-term solution for the refugee crisis

Empower our competitivity

manage to reduce their CO_2 emissions quickly, drastically and keep them low, and if we can show that we can live well, without destroying the biosphere, we can hope that other large regions of the world will also accelerate their fight against climate change. President Biden could adopt a pact quite similar to the European one.

- The Climate Pact would make it possible to create jobs, massively. In a Europe undermined by unemployment and job insecurity, this is obviously an essential point.
- The Climate Pact would make it possible to decrease speculation. If monetary creation goes to the real economy and employment, instead of going essentially to speculation, we will calm the financial markets. Similarly, if 5 per cent of profits are systematically oriented towards investment (private and public) instead of going increasingly towards dividends, we will reduce the resources of speculators. We will still be very far from the 'thirds' promised by Nicolas Sarkozy in February 2009 ('one third of profits should go to investment, one third to employees through bonuses and one third to shareholders'), but we will be a little closer.
- The Climate Pact would make it possible to strengthen our research in this strategic area. If Europe sets itself ambitious research targets and devotes €10 billion per year to achieving them, in addition to existing budgets, we could create a European research community in this area, which would soon be open to scientists from all over the world, in a wide variety of fields (the car of the future, the computer of the future, new materials, new ways of storing energy, agronomic research to achieve high yields with reduced inputs, and so on).
- The Climate Pact would make it possible to improve purchasing power. European Commission studies show that each household could save between €800 and €1,000 per year if we make the ecological transition: saving on heating and transport; saving on costs of healthcare; not to mention living better, breathing better and being better in ways that

mere accountancy cannot measure. In the early years, a large portion of the savings would be used to finance the transition (to repay the 0 per cent loans), but they will constitute purchasing power (or the power to live better) for all.

- The Climate Pact would make it possible to provide a substantive response to the refugee crisis. We all know that the number of refugees can only increase dramatically if Africa and other parts of the world are afflicted by climate disruption. By allowing Africa to finance its ecological transition and adaptation policies to whatever global warming we have not been able to avoid, the Climate Pact provides a substantive response to this question. Massively creating quality jobs on both sides of the Mediterranean and laying the groundwork for a new partnership between Europe and Africa, the Climate Pact must also help us to change our feelings about other people, and so to change the terms of the debate. 'One planet, one family' and 'Africa and Europe, two continents but one future' – these slogans are more positive than those of the populists who play on the fear of refugees.
- The Climate Pact would make it possible to strengthen our competitiveness. Vladimir Putin and the King of Saudi Arabia are raising the price of energy from oil, so it is fundamental that Europe drops its dependency on these imported sources of energy. On average, over the last ten years, France's annual trade deficit was €52.1 billion. The energy deficit (gas, oil and uranium) was, on average, €51.5 billion, so almost all of our deficit is a result of the energy we import! To halve our energy consumption and, for the remaining energy needs, use *Made in Europe* energy, whose price does not depend on foreign powers and that provides employment for Europeans, would be very good.
- Beyond all the issues already mentioned, it seems to us that the adoption of this Climate Pact would also make it possible to revive the European project. Since mid-2017, Angela Merkel and Emmanuel Macron have spoken about

relaunching the European project. There is an urgent need to give a clearer vision and content to the European project. The adoption of this Climate Pact would give Europe a new impetus – a new enthusiasm.

NINE

Save the Climate and Save Europe? It Is Now or Never!

> World peace cannot be preserved without creative efforts to match the dangers that threaten it.
>
> Robert Schuman, Declaration of 9 May 1950

While Europe, and the world, seem to be slowly but surely moving towards chaos, we should remember the words of Robert Schuman on 9 May 1950, when he decided with Konrad Adenauer to create the coal–steel union. To make war impossible in Europe, 'France and Germany offer their neighbours limited but decisive action', said Schuman modestly; however, the press was not mistaken, announcing the next day 'A revolutionary initiative.'[1]

Today, what is the 'limited but decisive action' that will allow us to break with the stifling processes at work in Europe and all over the planet? What revolutionary *decision* could put out, in a few months, a major part of the fires that are starting and could allow people to build, in a few years, a better future? The European Climate Pact presented in this book is a potentially powerful lever, capable of resolving many of the crises that are undermining the world today. Who can deny the need to invest heavily to help all European states, and all families, businesses, communities and associations in Europe, to finance their share of the work to achieve 'zero net emissions' of greenhouse gases by 2050?

Who can also deny the need for massive aid to the countries of the South? Historically, since the beginning of the Industrial Revolution, Europe has been the largest producer of greenhouse gases, but it is still relatively protected from the consequences of climate change, while Southern countries, which are only marginally responsible for the problem, are seriously affected. The point here is not to blame ourselves, but only to tell the truth, and to be aware, as Jean-Pierre Raffarin said in 2003, that Europe and Africa have 'common but differentiated responsibilities'.[2] Millions of people leave their homes and risk their lives crossing the Mediterranean, not lightly, but because they cannot live decently in the places where they were born.

We all know that the migrant crisis, which causes so many human tragedies and disrupts Europe, is only a small foretaste of what awaits us in 10 or 20 years if we are not able to help African countries fight against the lack of employment, and against climate deregulation, while adapting to the global warming that cannot be avoided. In 1948, when the US decided on the Marshall Plan to help Europe recover, 80 per cent of the funds coming into Europe were donations, not loans. It is urgent that Europe now offers this type of support to the countries south of the Mediterranean. In July 2017, Angela Merkel received a number of African heads of state in Berlin, and all agreed on the need for massive support to Africa to enable the best use of all the resources available (human and material) to accelerate economic development without having to resort to carbon-based energy sources. Years later, the question of financing this new 'Marshall Plan' remains unresolved, while the Sahel is facing the worst drought in 1,600 years.

'The Euro-Mediterranean region is a hotspot for the issue of climate change', underlines Jorge Borrego, Secretary General of the Union for the Mediterranean:

> The 43 Member States of the Union for the Mediterranean face considerable challenges related to the adverse effects

> of climate change. This is a hotspot on the IPCC maps.... If the growth of energy needs in the Southern Mediterranean continues on a trend of +10% per year, it will be necessary to do everything so that these needs are met by renewable sources. Beyond the Mediterranean, it is a real Europe–Mediterranean–Africa strategy that must be promoted to respond effectively to these challenges and opportunities. Climate change is a great opportunity to achieve more sustainable, inclusive and equitable development models.[3]

Moreover, Kako Nubukpo, former Minister of Forward Planning of Togo, leaves no doubt: 'I start from the fact that the development of Africa is the security of Europe. We have a community of destiny. People expect an inclusive vision of shared prosperity between the EU and Africa.'[4] Employment needs to be created on a massive scale – useful, sustainable and non-relocatable jobs – on both sides of the Mediterranean, from Copenhagen to Cape Town, via Venice and Bamako. Is this not a good way of greatly reducing migratory flows (both economic and climate migrants), as well as halting the desperation caused by unemployment and insecurity in all our countries?

A meeting of minds, the like of which has never been seen before

The project for a European Climate Finance Pact has only been on the table for a few months, but it brings together a growing number of supporters, from 12 countries and from all backgrounds: Laurence Parisot, former President of the Mouvement des Entreprises de France (Movement of the Enterprises of France [MEDEF]); Rudy De Leeuw, President of the European Confederation of Trade Unions; Pedro Sanchez, the Spanish Prime Minister; Jean-Pierre Raffarin and Jean-Marc Ayrault, former Prime Ministers of France, under Jacques Chirac and François Hollande; Monsignor Bruno-Marie Duffé, who works on these issues with Pope Francis; Guy Arcizet, former Grand Master of the Grand Orient; Prince Albert II of Monaco; Benoît de Ruffray, the Chief Executive Officer of Eiffage; Christophe Robert of the Abbé Pierre Foundation, the President of the European Affairs Committee of the National Assembly in Paris and the President of the European Economic and Social Committee; Christian Estrosi, the Mayor of Nice; Pascal Lamy, the former Director-General of the World Trade Organization (WTO); Franziska Brantner, a Member of the Bundestag known for her criticism of the WTO; Karl Falkenberg, former Director-General of Environment at the European Commission; Dany Cohn-Bendit; Enrico Letta, former President of the European Council; Olivier De Schutter, member of the UN Committee on Economic, Social and Cultural Rights; the Association of Rural Mayors of France; the Association of Small Towns; Alternatiba; France Nature Environnement; and a large number of other associations and thousands of citizens, from 16 to 96 years old.

At the headquarters of the EIB, a leader recalls: "In 1989, after the fall of the Berlin Wall, it took six months for François Mitterrand and Helmut Kohl to create a bank to finance the transition of the former Soviet countries." A European budget financed by a tax on profits has been proposed for the last 30 years, and everything is ready, technically, to do it. Announcing clearly that this budget will be financed from shareholder profits (and not simply by taxes on all citizens) and that it will create massive employment by financing the fight against climate change (a subject that concerns all member states) will undoubtedly contribute to obtaining support from a large number of countries. The Netherlands, for example, is hostile to the plan to strengthen Europe via the banking union proposed by Emmanuel Macron, but it is the country of Europe that is potentially most affected by climate change. It is no surprise that the 12 largest cities of the country have just elected ecologist mayors: if the sea rises by two metres, more than half of the country risks being flooded. The Netherlands, Sweden and Denmark, who are hostile to the agreement signed on 19 June 2018 by Angela Merkel and Emmanuel Macron, could be powerful allies to create a Europe that puts finance at the service of the climate.

Summit of 13 and 14 December 2018

Angela Merkel reminded people on 19 June 2018 that the renewal of Europe could no longer wait. She reiterated that it is this year, at the Summit of Heads of State and Government scheduled for 13 and 14 December in Brussels, when major decisions must be taken, and that next year, it will be necessary to ratify the new treaties (or modified treaties).

Merkel is right: *it is now or never!* Action must be taken, and everyone recognizes that Europe must acquire a new and greater vision that inspires its citizens. The peoples of Europe must no longer be tossed from one crisis to another, from one promise to another. 'Vote yes on the Maastricht Treaty, and we'll get to work on social Europe right afterwards', declared Jacques Delors in 1992. 'Fighting fiscal dumping and creating a European tax on profits? This is the next job for the Europeans', Jacques Delors explained a week before the 2005 referendum. However, the UK did not want a European tax and fiscal dumping has continued.

'If the Germans had had a referendum on the Maastricht Treaty, they would have voted NO, like the French', said a spokesperson of the Christian Democratic Union of Germany (CDU) in late 2005. 'We must hear the concerns of the people. We need to negotiate a European social protocol', said Angela Merkel in 2007 before all the European heads of state gathered in Berlin. Jacques Delors approved this idea, but Nicolas Sarkozy preferred 'go fast' and concluded a mini-treaty that had no social dimension. How can one not understand, therefore, the feeling shared by many Europeans of being let down?

'Our home is burning and we are looking elsewhere', said Jacques Chirac in 2002, calling on Europe to build 'an alliance of civilizations' with the countries of the South. More than 15 years have passed and we have not built anything solid. All indicators are red on the climate front. The number of climate refugees has never been greater. The populists fan fears, and our civilization is at risk of being ruined.

A Europe of joblessness and climate deregulation? So much time has been lost for these two key issues! Unfulfilled promises have fuelled the dissatisfaction of the peoples of Europe. What is happening in Italy and Germany shows that Europe is on the brink of a knockout. It is urgent that European leaders decide on *limited but decisive action*, concrete action that will have a visible, rapid and decisive impact on the lives of the millions of people who no longer believe in the European project. These millions of people feel a certain loss of dignity – that they count for nothing and suffer in their daily lives – and are sometimes tempted to find a scapegoat to blame for the harm done to them.

For all those who have no work or are in precarious jobs, Europe must quickly demonstrate that it is at their service. We need to prove wrong the people who think that Europe is incapable of reform because it is in the hands of lobbies and at the service of the banks.

The next European treaty must provide a complete response to the crisis, as the philosopher Cynthia Fleury said: 'We separate crises, but of course they are the same: today there can be no social justice without environmental justice. There can be no social contract without preserving ecosystem services.'

Yes, *it is now or never*. In 1950, it took two weeks for Schuman and Adenauer to decide on the coal–steel union – only two weeks, after decades of war and accumulated hatred! The EU was born with this economic community; it can be reborn with a climate and employment treaty.

In 1933, it took a few months for Roosevelt to create a federal tax. In 1941–42, it took him a few weeks to organize a metamorphosis of the US economy that won the war against Japan and Nazi Germany. When there is a very strong political will, it is possible to overcome, in a few months, obstacles that seemed insurmountable.

While Donald Trump wanted to reduce the world order built in 1945 to rubble, at the risk of causing chaos, it is urgent that Europe recovers its unity and injects into international relations a big dose of intelligence and fraternity, not just in speeches and statements, but in acts and budgets. 'We have a unique but small window for some key decisions', said Jean-Marc Ayrault, former French Prime Minister:

> But the context is difficult. However, we have an asset: an American President, who refutes the multilateral vision of the world; this is a good reason for Europe to assert itself as a political power. May the next EU treaties reflect a true understanding of the complexity of the world, a vision that gives meaning to the complexity.[5]

Is this transformation possible?

Is this 'great leap forward' that we are pushing possible? Yes. As Hölderlin said: 'But where the danger is, also grows the saving power.' In their heart of hearts, are a growing number of our leaders becoming aware that the current dominant politico-economic system is heading for collapse? They often have the greatest difficulty in confessing this publicly and of realizing what this means. On 9 October 2017, when he stepped down as Minister of the Economy in the Merkel government to take on the presidency of the Bundestag (the German National Assembly), Wolfgang Schaüble gave a major interview to the *Financial Times*. And what did he say? Did he summarize the glorious record of his ministerial action? Not at all! He took the opportunity to sound the alarm: 'Wolfgang Schaüble alerts the financial community on the risk of a future financial crisis', summarized the editor of the *Financial Times*: 'He denounces the risks posed by the runaway global debt – both at the state level and for private actors (companies and households) – and he underlines the threat of new bubbles, which have formed because of trillions of dollars created by central banks.'

Is the ECB deaf to these criticisms? Will it block the changes we are working for here? No, not at all. There is a growing feeling that Mario Draghi and several other ECB leaders are torn between their duty to the institution and their desire to ask governments to change the rules of the game.

Mario Draghi regularly makes it clear that he is not blind and that he is aware of the growing risks. However, he is faced with a dilemma: stop the ECB's very generous monetary policy and take the risk of Europe falling back into recession, or continue it and fuel financial bubbles. From time to time, he puts aside his reserve and hints that he would very much like governments to help him out of this dilemma at last, using the money available at 0 per cent interest rates for public investments that would boost the economy much more sustainably than the money does at the moment. Green investments? Yes, why not!

Just recently, Mario Draghi has also revived the debate on the need for a 'European tax instrument' to supply a budget that would allow Europe to better prepare itself for a new financial crisis. 'We need an additional fiscal instrument to maintain convergence during major shocks without having to overload monetary policy', Draghi said on 11 May 2018 in Florence. A 'European fiscal instrument'? A federal income tax, for example? Good idea! However, are we going to wait for the next crisis to erupt to put it in place, or are we able to anticipate a bit and strengthen our economies before the next crisis?

'A European budget? But the Bundesbank will say "no"', say some people. It has indeed long been a taboo subject in the corridors of the 'Buba', but the recent declarations of Jens Weidmann, the very powerful President of the Bundesbank, often presented as the next president of the ECB,[6] should be read attentively. In almost all his statements since the election of Emmanuel Macron, Weidmann blows hot and cold. He first explains that 'the debate initiated by Emmanuel Macron is interesting, but the common budget he seeks is not an objective in itself'; however, a few lines later, Weidmann says:

> We must do things in the right order: before talking about a common budget, we must discuss which functions are better served at European level than at the national level. What are the areas which affect us all that may require a transfer of responsibility/skills (and a budget) to the European level, such as environmental preservation or border protection?

Jens Weidmann is, therefore, a priori, hostile to an EU budget of several hundreds of billions of euros but is prepared to discuss special cases. This book describes a crucial area that requires just this kind of transfer of 'responsibility/skills' and funds.

A transformative treaty, adopted by a pan-European referendum?

As we described earlier, we are working for a European treaty to ensure the stability of funding for the ecological transition. When Nicolas Hulot explained the reasons for his resignation as Minister for the Ecological and Social Transition on 28 August 2018, he mentioned that the budget he required to promote the insulation of buildings in France had been halved. This kind of instability in policies discourages everyone involved in major transformations. In *Laudato si*, the Papal encyclical dedicated to caring for our common home, Pope Francis insists on this point: 'Continuity is indispensable because policies to protect the climate and the environment cannot change whenever there is a new government.'

To counter climate change, each of us must change our personal way of life, but these individual changes will never

Box 9.1: The ECB advocates for a rise in public investments

Growth in the Euro area economy would soon benefit from higher public investment relative to the current low levels, as the low inflation environment and accommodative favorable conditions for an increase in spending, said the European Central Bank last Monday.

The ECB has long advocated for eurozone governments to invest more in addition to its ultra-accommodative monetary policy designed to support growth and inflation.

Since the sovereign debt crisis in the euro area, public investment has fallen to record lows, weighing on potential growth.

'The ECB can not support growth alone.... An increase in public investment has a positive effect on demand', writes the ECB in its economic bulletin. 'This finding calls for increased public investment in the current low inflation environment.'

Reuters (2016) 21 March, Frankfurt.

be enough if the collective rules do not change.[7] As Prince Albert II reminds us:

> we are now absolutely certain that warming is mainly due to human activities. Thanks to science, we understand its undeniable consequences. Year after year, these facts have been dissected, substantiated, specified. I doubt that it is still possible to ignore them for long. *The responsibility of political decision-makers is essential: it is up to them today to build the legal framework which will make it possible to change our development model.*[8]

A European treaty would guarantee the stability of this new legal framework. As Mireille Delmas-Marty[9] says clearly, this treaty would be a 'catalysing treaty', with an effect like an electric shock. It would make possible things that seemed impossible, or that would happen far too slowly. Moreover, the catalytic effect will be even stronger if this treaty is approved by a pan-European referendum, held on the same day in all European countries.

A pan-European referendum? The idea is not ours; it comes from Jürgen Habermas, the great German philosopher, who suggested it publicly in 2007, and again in 2012. Initially, German political leaders were very reserved (there has been no referendum in Germany since the fall of the Nazi regime); then, a majority of them accepted the idea, as Europe can no longer be built behind people's backs. If Europe is at the service of the people, it should not be afraid to give them the floor to approve future treaties.

'A real pan-European referendum, i.e. a referendum organised on the same day in all countries, is tantamount to "denationalising" the question posed, and thus avoiding having the question attached to national political issues, often superficial', says Mireille Delmas-Marty. This overcomes the problems of many referenda. The question is no longer, 'Are you for or against Emmanuel Macron, Angela Merkel or Pedro Sanchez?', but rather, 'Do you want Europe to put finance at the service of climate and employment?'

To provoke a debate at the European level is also to create 'co-citizenship' for more than 500 million people, and so helps in the creation of a European identity. It also anticipates future stages, once the treaty has been accepted: the campaign 'Yes to the Climate Pact', which will no doubt be intense in all countries, will lead to multiparty and transnational meetings, which can then help launch cooperation and concrete projects for the subsequent implementation of the Climate Pact on the ground.

Yes, it will be an exciting debate. In all families, in all villages and towns of Europe, in all companies and associations, at school gates, in churches, synagogues and mosques, and in theatres, cinemas, gymnasiums, universities, cafes and trains, we will discuss climate issues. Millions of people will realize at the same time the seriousness of the situation and all the opportunities that this crisis provides for a new Europe. Yes, we can live better by consuming much less energy! Yes, there is joy in deciding our future together – and in changing the course of history!

What if, in spite of everything, there is a vote against the Climate Pact in a few countries? Or, if one nation says 'no' to a project that offers money to fund massive job creation by protecting the future of its children? Then it will be necessary to develop 'reinforced cooperation', as already foreseen by the current treaties: a whole continent could not allow itself to be blocked by the decision of a minority of countries. We could make a start in 12, 13, 14 or 15 countries – and maybe, a few years later, the initially reluctant countries would join the pioneers.

On the same day when the citizens are called upon to vote in a referendum on the European treaty, we could also vote in each country for a national plan for the practical implementation of the ecological transition – an 'action plan' that will define the solutions and new standards specific to each country. Of course, from Lapland to Cyprus, from Tallinn to Lisbon, people's needs and resources differ. Each country will be invited to consider the application of the Climate Pact. The referendum would be an opportunity to obtain the agreement of the people on the new rules of the game that will accelerate the transition: 'In 20 years, all buildings will have to be insulated. In 20 years, no car will be allowed on the roads if it produces more than x kilogrammes of CO_2 per 100 kilometres' and so on. If these decisions are decided by referendum, no one could then oppose their implementation since the new standards would be the result of a democratic choice, and the European Climate Pact would allow every country to finance the projects necessary to reduce our impact on the climate. The choices the Greeks make would not be the same as those of the Irish or the Romanians, but what a brainstorming there would be all over the continent! The climate issue would finally be a priority not only for some ministers, some scientists or some activists, but for tens of millions of people who will have decided to vote yes, to change Europe, change their lifestyles and 'save the climate'.

Yes, if we really follow the European news, we can see that the lines are moving: the old world is dying and some leaders are changing their tune. However, they are not yet changing their actions (Alas!), or are but a thousand times too slowly. This is the problem. As Nicolas Hulot says, because of this inertia and the weight of certain lobbies, 'the planet is falling into a tragedy'. To overcome this inertia and to defeat certain lobbies, it is fundamental that as many citizens as possible launch a dynamic that will force politicians to be ambitious, to be daring.

Conclusion: Creating a New Development Model

The scenario of collapse is easy to imagine: a financial crisis that causes a surge of unemployment and insecurity; populists come to power in many countries; Europe falls apart; and the fight against climate deregulation is postponed – until it is too late! This scenario is entirely plausible. It is even most likely, alas, given the inertia of our leaders over recent decades and the COVID-19 crisis. Fortunately, the story is never written in advance.

Stéphane Hessel is no longer with us, but he supported this Climate Finance Pact. Stéphane always said that we must refuse to be discouraged. On all continents, we have seen situations that seemed completely hopeless, but in a few months, the will of the citizens changed the situation radically: 'We experienced apartheid and the end of apartheid', he said: 'We had the Berlin wall and saw the end of the wall…. It's up to us, the citizens, to decide on our future.'

Yes, it is up to us to decide. It is time to wake up. It is time to raise our heads. It is time to pull together. It is time to convince our fellow citizens to come together, to push our leaders to be brave. *What if this climate crisis gave humanity a great opportunity?* It is indeed the first time that the whole of humanity is facing a major challenge together – pick up the gauntlet or die – and to meet this challenge, we must radically change our way of living together. Our solidarity must be expressed at all levels: from our neighbourhood, our work colleagues and our family to the universal, and from our region to our continent. All humans are concerned.

Olivier de Schutter, the new Special Rapporteur for the Right to Food at the United Nations, is correct: it is time to stop small steps.[1] Nicolas Hulot says the same thing.[2] It is time to declare war on climate change (yes, *declare war!*) because it is the only war we absolutely have to win today, the only war in which there will be no victims, but which will avoid millions of deaths, and the only war that can unite peoples instead of dividing them.

It is the only war that can bring together the US and Turkey, Iran and Israel, those who believe in heaven and those who do not, those who have children and those who do not, and those who live in the posh suburbs of Los Angeles and those who live in the ghettos of Bangkok. The choice seems so obvious! On the one hand, chaos, famines and endless conflict; on the other, a promise of living better – humanity reconciled with nature and reconciled with ourselves. We need a Europe that recovers its values and opens the way to a change of civilization.

To win this essential battle, will you join us?

See the Climate Finance Pact 2019, available at: https://agirpourleclimat.net/en/

Notes

How We Can Win the Battle

1. Minister of State, Ministre de la Transition écologique et solidaire, speaking at the conclusion to the meeting on the 'Climate and Finance Pact'.

Chapter 1

1. Intergovernmental Panel on Climate Change (IPCC) (2012) 'Managing the risks of extreme events and disasters to advance climate change adaptation', https://www.ipcc.ch/report/managing-the-risks-of-extreme-events-and-disasters-to-advance-climate-change-adaptation/
2. The figure comprises total expenses incurred by insurers, local governments and citizens.
3. AfricaNews with Reuters.
4. *La Tribune* with AFP, 13 February 2018.
5. The summer of 2003 (June, July and August) was 3.2 °C hotter than the long-term average.
6. Bador, M., Terray, L., Boé, J., Somot, S., Alias, A., Gibelin, A.-L. and Dubuisson, B. (2017) *Environment Research Letters*, 12. https://iopscience.iop.org/article/10.1088/1748-9326/aa751c
7. It is difficult to speak of Africa in the singular, so large is the continent, and so varied the situations on it. A book could be devoted to the impact of global warming on the different regions of Africa, and another on the different regions of America or Asia.
8. Kang, S. and Eltahir, E.A.B. (2018) *Nature Communication*.

Chapter 2

1. The debate on the *causes* of global warming is 'done' in the eyes of many, but there was still discussion (and therefore further research needed) on: its *intensity*, +1.5 °C or +4.5 °C for a doubling of the concentration of CO_2; how long the warming will take; what impact it will have on the level of the oceans; and what impact it will have on human nutrition. For the full *New York Times Magazine* article, see: www.nytimes.com/interactive/2018/08/01/magazine/climate-change-losing-earth.html
2. This extract is from an excellent article: Chambonnier, H. (2015) *Le Télégramme*, 23 November.
3. On 10 August 2018, following the lawsuit of Dewayne Johnson, a gardener who suffers from cancer, Monsanto was fined US$289 million for

continuing to sell a potentially carcinogenic herbicide, without informing users of the dangerous nature of the product, when the company was in full knowledge of the problem.

4 Not all Americans are like Donald Trump! In 1958, there were already some good theoretical arguments to justify interest in these measures.

5 Allègre, C. and de Montvalon, D. (2010) *L'Imposture Climatique ou La Fausse Écologie, conversations avec Dominique de Montvalon*, Paris: Plon.

6 CO_2 acts in ways that remind us of hormones in the human body; for example, even for an adult of 80 or 90 kilograms, a few micrograms of excess thyroid hormone are enough to disrupt a good deal of the person's physiology. Similarly, 'just' a few tens of ppm more CO_2 in the entire atmosphere is enough to disrupt the planet's climate, in the same way that a thin glass plate a few millimetres thick keeps the huge greenhouses of a plant collection warm.

7 There is much more vegetation in the Northern Hemisphere than in the Southern Hemisphere, which explains why the influence of the seasons is so strong in the measures from Hawaii.

8 Jouzel et al (1987) *Nature*, October. Barnola et al (1987) *Nature*, October. Genthon et al (1987) *Nature*, October.

9 Chappellaz et al (1990) *Nature*.

10 Petit et al (1999) *Nature*. Siegenthaler et al (2005) *Science*. Luthi et al (2008) *Nature*. Loulergue et al (2008) *Nature*.

11 IPCC (2013) 5th Report, Group I.

12 To collect ice samples, climatologists often spend several weeks in Antarctica, far from their families and sometimes in very harsh conditions.

13 See the study published by the Proceedings of the American Academy of Sciences (PNAS) on 15 August 2017, available at: https://doi.org/10.1073/pnas.1709070114

Chapter 3

1 Roger, S. and Foucart, S. (2017) *Le Monde*, 13 November.

2 Deforestation is especially visible in the South, but we all know that it is mainly to feed the countries of the North that forests are destroyed, among other things, to sow soya.

3 Quoted in *El Païs* (2018) 12 August.

4 See Will Steffen et al (2018) 'Trajectories of the Earth System in the Anthropocene', *PNAS*, July.

Chapter 4

1. UN Environment Programme (2017) *The Emissions Gap Report 2017*, Nairobi: UN Environment Programme.
2. A warming of 1.5–2 °C above the pre-industrial climate, that is, 0.5–1 °C above 2010.
3. Goal announced in the 2015 French Parliamentary Act for 'Energy Transition for Green Growth'.

Chapter 5

1. This was the 2050 target, based on 1990 emissions, in an announcement made by Jean-Pierre Raffarin on 19 February 2003 during the opening of the plenary assembly of the IPCC in Paris.
2. The quote is from discussions at the SFP, initiated by Michel Spiro.
3. IPCC (2014) 5th Report, Group 3.
4. Those who have already waited for a bus or train for an hour under the cold rain prefer taking their car, even if it means being stuck in traffic (and massively producing CO_2) – at least, it is warm.
5. When transformed into renewable energy, farm waste becomes a source of revenue for farmers.
6. To produce one cubic metre of concrete, 398 kg of CO_2 are emitted, while a building built in wood will stock 425 kg of CO_2 per cubic metre of wall (for its entire life). We now know how to build beautiful apartments entirely of wood (see 'Building differently', available at: https://agirpourleclimat.net/en/).
7. Energy use attributed to the internet has clearly increased during the last decade.
8. If we're giving ourselves 30 years to insulate *all* of the buildings, we can be smart about it and start today with those that use the most energy.
9. When we walk or bike, we burn fats and we activate our leg muscles, which, in turn, activates the lymphatic system and consequently the immune system, and we also add fewer fine particles to the air.

Chapter 6

1. For the first time in US history, the number of married women in employment exceeded that of spinsters.
2. Yuri Gagarin on 12 April 1961.

Chapter 7

1. The melting ice is adding more and more water to the oceans, where sea levels are rising anyway because of the expansion of the water caused by the increase in temperature.
2. People report suffering from the cold in 3.5 million households in France – 1.6 million if we consider only those whose income is in the lowest 30 per cent. Moreover, a study in 2008 by the governmental Agence de l'environnement et de la maîtrise de l'énergie (Agency for Ecological Transition [ADEME]) shows that the 5 million lowest income households spend 15 per cent of their income on energy expenditure (housing and transport), compared with only 6 per cent for the better off. If we take into account the various indicators studied by the Observatoire national de la précarité énergétique (National Observatory of Energy Precariousness), 5.1 million households (12 million individuals) are in a situation of fuel poverty, with sometimes very serious consequences for health (in particular, children's health).
3. The High-Level Expert Group on Sustainable Finance (set up at the European level to consider how to finance the transition and particularly the contribution of private finance in this fight) assesses the cumulative investments necessary to achieve the objectives at about €11,200 billion between 2021 and 2030, and the investment deficit at nearly €2,000 billion (see 'Financing a sustainable economy – interim report by the High-Level Expert Group on Sustainable Finance', July 2017).
4. In 2010, after a complaint lodged against the Federal Reserve by two Bloomberg journalists, a judge gave them access to the archives of the US Central Bank. After poring over 20,000 pages of various documents, they found that the Federal Reserve secretly lent troubled banks the sum of US$1,200 billion at the incredibly low rate – at the time – of 0.01 per cent. Michel Rocard and Pierre Larrouturou said: 'Why do states have to pay 600 times more than banks?' (Rocard, M. and Larrouturou, P. [2012] *Le Monde*, 2 January). The figures of money creation made public by the central banks are already colossal, but it would be naive to think that banks are completely transparent in all their actions.
5. Created in 1983 'in response to the international debt crisis' (*No comment!*), the IIF currently has more than 450 members (banks, hedge funds and other financial institutions), with offices in more than 70 countries on all continents.
6. US household debt reached US$15,300 billion as at 31 March 2018.
7. The stated figure does not count the US$15.6 trillion in financial sector debt (see Federal Reserve [2018] 'Flow of funds', June).

[8] Source: Bank of China, August 13, 2018. New loans distributed by Chinese banks accounted for a record 13,530 billion yuan in 2017 (just over $2 trillion), but this figure will be exceeded in 2018. January 2018 was the month of all records: 2,900 billion yuan of new credits distributed in one month only! And the total during the first six months of 2018 reached 9,300 billion. The Chinese government wants to 'calm credit', but the total value of new loans increased by 13% in just a year. With 1,450 billion yuan loaned in July 2018, there is still no real lull.

[9] Patrick Artus is a director of studies at Caisse des Dépôts, CDC Natixis, and Professor of Economics at the École Polytechnique. See, in particular, the analysis proposed by Christian Chavagneux of Patrick Artus' book *Les Incendiaires: Central Banks Overwhelmed by Globalization*, in *Alternatives économiques*, October 2007.

[10] 'In this book, I hope not to shock, especially those who express themselves with elegance and moderation', writes J.-M. Naulot: 'But given the gravity of the crisis we are going through and the one that is on the way, I felt the need to speak directly, without double talk' (Naulot, J.-M. (2017) *Éviter l'effondrement*, Seuil Editions).

[11] Interviewed by De Filippis, V. (2017) *Libération*, 7 February.

[12] Le Boucher, E. (2018) 'La terrible fuite en avant de la dette mondiale', *Les Echos*, 13 April.

[13] Nicolas Barré is quoting the IMF here.

Chapter 8

[1] Grandjean, A. and Giraud, G. (2018) *Alternatives Economiques*, September.

[2] Our parliaments should create structures for dialogue between parliamentarians, experts, NGO officials, trade unionists and employers' leaders.

[3] Philippe Maystadt passed away in December 2017, suffering from a lung disease against which doctors could no longer do anything. He showed incredible courage: a few days before the end, he sent an email to his friend Michel Barnier to invite him to commit himself to the Climate Pact. Placed at the centre of the political spectrum, Philippe Maystadt was a great finance minister of Belgium: he put the country's finances in order while always caring for the poorest and for the generations to come. One cannot pay too much tribute to his intelligence, his kindness and his generosity.

[4] They support our project as citizens, but at this stage, they obviously do not commit the EIB or the ECB.

[5] This figure of 38% is the sum of the Federal tax created by Roosevelt (35%) and income tax 'residues' in almost all states (3% on average).

6 The 30 per cent share paid by the member states corresponds, roughly, to what is already available in several countries in the form of a tax credits. Furthermore, the states would be helped because when these works are done, they collect more VAT and social contributions. However, the current system is often complicated, unstable (the rules change from one year to another) and partial (the tax credit corresponds to 30 per cent of the price of the materials but does not necessarily cover the labour costs), and obliges the individuals or the companies to advance the payment while waiting to benefit from a tax cut the following year. To speed up the process, it is necessary to set up a simple, robust, stable, complete and immediately effective system.

7 On 8 March 2018, the European Commission unveiled its strategy to encourage private finance to become more committed to climate change. Following the recommendations of the High-Level Expert Group on Sustainable Finance, the European Commission wants to 'allow both investors and individuals to make positive choices so that their money can be used more responsibly'. For this, the European Commission wants to establish a unified classification system to define what is sustainable and what is not. It would then create labels for green financial products. The European Commission also wants to clarify the requirement for asset managers and institutional investors to take sustainability considerations into account in the investment process, and to strengthen their reporting obligations. The European Commission's report now has to be translated into concrete terms. Just as we have strengthened (somewhat) the operating rules of banks after the crisis of 2008, we must quickly improve the rules applied to banks and insurance companies to stop investments that negatively affect the climate and to accelerate sustainable investments.

8 'The carrot before the stick': to drastically increase the carbon tax without having first given the majority the means to insulate their houses or adopt energy-efficient modes of transport would be counterproductive. However, there is no doubt about the usefulness of increasing the price of carbon in the long run. On this subject, see the publications of the working group led by Joseph Stiglitz, who proposes the establishment of a Carbon Pricing Corridor.

Chapter 9

1 *Le Figaro* (1950) 10 May.

2 This principle of common but differentiated responsibilities was included in the Rio Climate Convention (in 1992) and the Kyoto Protocol (in 1997). Putting aside the controversies aroused by a reductionist interpretation of a binary differentiation between the industrialized countries and

all the others, Mireille Delmas-Marty argues that such a principle will be indispensable in all areas, especially in the face of human migration, as long as the diversity of the world's legal systems excludes a universal common law.

[3] Speech at a meeting on the Climate Finance Pact, 15 March 2018 at UNESCO, Paris.

[4] Speech at a meeting on the Climate Finance Pact, 15 March 2018 at UNESCO, Paris.

[5] Intervention during the day of debate around the Climate Finance Pact, 15 March 2018 at UNESCO, Paris.

[6] See, in particular, *Les Echos* (2017) 3 November, as well as *Ouest France* (2018) 19 May.

[7] If it is important to switch off the tap while washing one's teeth, it is also important to invest enough in the water networks to avoid massive losses, which vary between 9 and 34 per cent across the cities of France.

[8] Speech by Prince Albert II, Oxford Union, 6 February 2017.

[9] Mireille Delmas-Marty was a member of the Institut Universitaire de France between 1992 and 2002 and Emeritus Professor at the Collège de France. She is one of the supporters of the European Climate Pact. Her excellent text, *Climate Risk: A Chance for Humanity*, shows the scale of changes needed and the fundamental values we must reaffirm. More than ever, we are all citizens of the world.

Conclusion

[1] See the article dated 3 May 2020, available at: https://news.un.org/fr/story/2020/05/1068012

[2] See page 11 of this book. Nicolas Hulot is the most popular public figure in France, being an activist for Nature, film director, broadcaster and France's Minister for the Environment May 2017–September 2018.

Index

D

I

J

K

L

Q

R